Blood Groups - More than Inheritance of Antigenic Substances

Edited by Kaneez Fatima Shad

Published in London, United Kingdom

IntechOpen

Supporting open minds since 2005

Blood Groups - More than Inheritance of Antigenic Substances
http://dx.doi.org/10.5772/intechopen.94700
Edited by Kaneez Fatima Shad

Contributors
Amr J. Halawani, Edith Christiane Christianne Bougouma, Alphonse Ouedraogo, Sodiomon Bienvenu Sirima, Ana Azanjac Arsic, Kaneez Fatima Shad, Sameer Saleem, Wissam Soubra, Dennis J. Cordato, Francisca Varpit, Vela Galama, Yetti Hernaningsih, Chima Akunwata, Williams Bitty Azachi, Kuschak Mathias Dakop, Parviz Lalezari, Behnaz Bayat

First published in London, United Kingdom, 2022 by IntechOpen
IntechOpen is the global imprint of INTECHOPEN LIMITED, registered in England and Wales, registration number: 11086078, 5 Princes Gate Court, London, SW7 2QJ, United Kingdom
Printed in Croatia

British Library Cataloguing-in-Publication Data
A catalogue record for this book is available from the British Library

Additional hard and PDF copies can be obtained from orders@intechopen.com

Blood Groups - More than Inheritance of Antigenic Substances
Edited by Kaneez Fatima Shad
p. cm.
Print ISBN 978-1-83969-902-3
Online ISBN 978-1-83969-903-0
eBook (PDF) ISBN 978-1-83969-904-7

For EU product safety concerns:
IN TECH d.o.o., Prolaz Marije Krucifikse Kozulić 3, 51000 Rijeka, Croatia, info@intechopen.com or visit our website at intechopen.com.

Meet the editor

Professor Kaneez Fatima Shad, an Australian neuroscientist with a medical background, obtained a Ph.D. from the Faculty of Medicine, University of New South Wales (UNSW), Australia, in 1994, followed by a postdoc at the Allegheny University of Health Sciences, Philadelphia, USA. She taught medical and biological sciences at various universities in Australia, the United States, United Arab Emirates, Bahrain, Pakistan, and Brunei. During this period, she was also engaged in research by obtaining local and international grants (a total of more than $3 million USD) and developing products such as a rapid diagnostic test for stroke and other vascular disorders (i.e., schizophrenia). She has published more than sixty-eight articles in refereed journals, edited nine books, authored ten book chapters, presented at more than ninety international conferences, and mentored thirty-four postgraduate students. She is an international mentor and a protocol development specialist.

Contents

Preface

Every individual is born with a set of antigenic substances inherited from their parents. These special molecules are mainly present on the membrane surface of red blood cells. These structures can be sugars, proteins, or a special set of enzymes that can make an individual prone to or protected against various infections and other diseases.

Besides the ABO blood group, there are more than thirty-seven other systems with more than 300 blood group antigens present in the human blood. These blood group structures not only perform a critical role in the function of cells but can also be used by viruses to gain access to cells.

This book includes nine chapters on different blood group antigens and their activities behind the scenes.

Chapter 1, "Are ABO Gene Alleles Responsible for Cardiovascular Diseases and Venous Thromboembolism, and Do They Play a Role in COVID?" by Kaneez Fatima Shad et al. discusses the association between ABO blood groups and cardiovascular diseases (CVD). The authors discuss how individuals with non-O blood group type are at increased risk of coronary heart disease, myocardial infarction, cerebral ischemic stroke, peripheral arterial disease, and venous thromboembolism (VTE) including deep vein thrombosis and pulmonary thromboembolism. The most likely mechanism in all cases is thought to be the increase in von Willebrand Factor (vWF) and factor VIII levels seen in patients with a non-O blood group. The increased levels in vWF and factor VIII among individuals with a non-O blood group who have contracted SARS-CoV-2 infection may result in an additive thrombophilic effect to that caused by the virus.

Chapter 2, "Human Blood" by Francisca Varpit and Vela Galama discusses the red cell component of human blood, specifically red cell membranes and the antigens found on them. This chapter briefly covers the genetic inheritance of common antigens, the important structural makeup of red cells, antibody response to these antigens, and their functions in health (physiological) and disease (pathophysiological). Some of these blood group antigens maintain the structural integrity of red blood cells and play a role in the cationic exchange between intracellular and extracellular molecules, without which red cells become disfigured resulting in a reduction in their survival and increased risk of hematological diseases. These antigens can also serve as transporters, adhesion molecules, and receptor proteins relaying chemical messages from cell to cell. However, these mechanisms have been exploited by infectious agents to gain entry into the human body causing disease.

Chapter 3, "ABO Blood Group and Thromboembolic Diseases" by Yetti Hernaningsih discusses the division between type O and non-O blood groups due to the presence of antigens on the cell surface and antibodies in plasma and their relationship in the onset of thromboembolic diseases in the non-O blood group due to higher levels of VWF and FVIII. The author further discusses how arterial thromboembolic events are often associated with cardiovascular diseases, and deep vein thrombosis (DVT) and pulmonary embolism (PE) with venous thromboembolism (VTE).

Chapter 4, "Duffy Antigens and Malaria: The African Experience" by Chima Akunwata describes the Duffy antigen/chemokine receptor (DARC), which is located on the surface of red blood cells. It is a receptor for pro-inflammatory chemokines and for the invasion of Plasmodium vivax into the red blood cells. A point mutation in the promoter region of the Duffy gene disrupts the binding of a transcription factor leading to a lack of expression of the antigen on the erythrocytes. This Duffy negative phenotype is found predominantly in the African population. This mutation is advantageous as individuals with the Fy(a-b-) phenotype are less susceptible to P. vivax malaria. It is believed that the absence of Duffy antigen in most Africans contributed to the resistance to P. vivax and by extension reduced the burden of malaria in these endemic areas.

Chapter 5, "Blood Groups: More than Inheritance of Antigenic Substances - Susceptibility to Some Diseases" by Williams Bitty Azachi and Kuschak Mathias Dakop links diseases such as infections, coagulopathy, coronary, malignancy, and malaria with ABO blood groups. The authors' findings reveal that the O blood group has a greater frequency of severe infections such as Escherichia coli and cholera, whereas blood group A is associated with the incidence of smallpox and some bacterial infections. These observations are principally based on the presence or absence of "H-like" and "A and B-like" antigen markers. Antigens A, B, and H are connected to N-glycans of vWF and reduce the half-life of this protein to 10 hours for group O and 25 hours for non-O groups. This explains the reason for increased levels of vWF in non-O individuals. This phenomenon raises the possibility of coagulopathy in non-O groups. Similarly, some tumors have A or A-like antigens, which explains the tendency of group A individuals to develop tumors. The A-like antigens of tumor cells are seen as foreign in B and O blood types with resultant immune responses.

Chapter 6, "RH Groups" by Amr J. Halawani describes a hemolytic disease of the newborn, now known as hemolytic disease of the fetus and newborn (HDFN), due to the presence of antigens in the RH blood group system. To date, fifty-six antigens have been recognized within the RH blood group system. The five main antigens are D, C, c, E, and e, and the antibodies to these antigens are implicated in hemolytic transfusion reactions (HTR) and HDFN.

Chapter 7, "Neutrophil-Specific Antigens: Immunobiology, Genetics, and Roles in Clinical Disorders" by Parviz Lalezari and Behnaz Bayat is about neutrophil-specific antigens (i.e., NA and NB), which are responsible for inducing febrile and potentially fatal lung injury and causing graft rejection in bone marrow transplantation. In clinical studies, it is also demonstrated that NB expression is highly elevated in polycythemia vera, and in some carcinomas, NB is expressed on the tumor cells. Neutrophil-specific antigens are examples of structures that beyond antigenicity have important biological activities and clinical implications.

Chapter 8, "The ABO Blood Group System and *Plasmodium falciparum* (*Pf*) Infection in Three Ethnic Groups Living in the Stable and Seasonal Malaria Transmission Areas of Burkina Faso (BF)" by Edith Christiane Bougouma et al. discusses the importance of blood group and its role in the onset of Plasmodium falciparum (Pf) infection. It also discusses the role of two specific genetic polymorphisms (rs8176719 and rs8176746) on such infections. The authors demonstrate that the Fulani ethnic group is not only less susceptible to Pf malaria infection but also has lower parasite densities when infected. The authors suggest that the origin, distribution, and relative proportion of ABO blood groups in humans may have been directly influenced by selective genetic pressure from P. falciparum infection.

Chapter 9, “ABO Blood Groups and Glioma” by Ana Azanjac Arsic is a modified version of the author’s previous publication “ABO blood groups and risk of glioma: A case-control study from Serbia” published in 2017. The aim of this case-control study was to investigate the possible association between the incidence of glioma and certain blood groups. Univariate binary logistic regression analysis revealed that individuals with group AB were at a 3.5-fold increased risk of developing glioma compared to individuals with other ABO blood groups. Mechanisms that explain the relationship between blood groups ABO and cancer risk are unclear. Several hypotheses have been proposed, including a modulatory role of blood group ABO antigens and regulation of the level of circulating proinflammatory and adhesion molecules, which are important for tumorigenesis by the blood group ABO system. One of the possible explanations for the role of the blood group ABO in the tumorigenesis process is the recent discovery of vWF, which is an important modulator of angiogenesis and apoptosis. The results of the study suggest that the blood group AB may be one of the hereditary factors that have an influence on the occurrence of glioma.

I would like to thank Author Service Manager Maja Bozicevic at IntechOpen for her hard work, constant support, and help.

This book is meant for a range of readers with different levels of interest in these incredible antigens, which are not only defining our blood groups but are also responsible for protecting us from the infection caused by bacteria and viruses.

Kaneez Fatima Shad
School of Life Sciences,
University of Technology Sydney,
Sydney, Australia

Chapter 1

Are ABO Gene Alleles Responsible for Cardiovascular Diseases and Venous Thromboembolism and Do They Play a Role in COVID?

Dennis J. Cordato, Wissam Soubra, Sameer Saleem and Kaneez Fatima Shad

Abstract

Cardiovascular diseases (CVD) including coronary heart disease and stroke are leading causes of death and disability globally. Studies of the association between ABO blood groups and CVD have consistently demonstrated an increased risk of coronary heart disease, myocardial infarction, cerebral ischaemic stroke, peripheral arterial disease and venous thromboembolism (VTE) including deep vein thrombosis and pulmonary thromboembolism in patients who possess a non-O blood group type. The most likely mechanism is thought to be the increase in von Willebrand Factor (vWF) and factor VIII levels seen in patients with a non-O blood group. Other postulated mechanisms include elevations in circulating inflammatory markers such as endothelial cell and platelet adhesion molecules in subjects with a non-O blood group. More recently, it has also been recognised that individuals with a non-O blood group type carry a higher risk of SARS-COV-2 infection and COVID-19 related complications. The increased levels in vWF and factor VIII amongst individuals with a non-O blood group who have contracted SARS-CoV-2 infection may result in an additive thrombophilic effect to that caused by the SARS-CoV-2 virus. Another postulated mechanism is that individuals with an O-blood group are protected by anti-A and B antibodies which possibly inhibit the binding of the SARS-CoV-2 spike protein to lung epithelium angiotensin converting enzyme-2 receptors. There are over 35 minor blood groups on red blood cells, some of which such as Kidd, Lewis and Duffy have been associated with CVD either alone or in combination with a non-O blood group allele(s). However, their role in SARS-CoV-2 infection and mechanism of action for an association with CVD remain unknown. This review explores the relationship between ABO and minor blood groups with CVD and VTE, with a focus on potential mechanisms underlying this relationship and the potential role of ABO blood group types in COVID.

Keywords: cardiovascular diseases, venous thromboembolism, ABO blood group, von Willebrand Factor, COVID

1. Introduction

Cardiovascular diseases (CVD) including coronary heart disease and stroke are the global leading cause of death and a major contributor to disability [1].

Traditional modifiable risk factors include hypertension, dyslipidaemia, diabetes mellitus, current smoking, obesity and physical inactivity and non-modifiable risk factors include age, gender, family history and ethnic background [2]. Amongst the non-modifiable risk factors, genetic variations in conjunction with traditional modifiable risk factors may significantly influence the trajectory of an individual's CVD risk [3]. Studies on the association between ABO blood group and CVD have consistently demonstrated that possession of the O blood group, the most common phenotype in most populations [4], confers protection against an individual developing a cardiovascular event [5–11]. The A and B blood groups are most frequently seen in Caucasian and Asian populations, respectively [4]. However, the magnitude of the association between CVD and ABO blood grouping across different ethnic populations is controversial in part due to the higher population attributable risk of traditional modifiable vascular risk factors [9, 10, 12].

There is also a well-documented interaction between ABO blood group and venous thromboembolism (VTE) [7, 13–15]. The A2 blood subgroup, which is less common than A1 and rare in Asian populations [5], has been reported to be associated with a modest VTE risk (Odds Ratio 1.2) whereas the A1 and B subgroups confer a 1.8-fold increased risk [7]. The non-O blood groups are associated with ~25–30% higher plasma levels of factor VIII and von Willebrand Factor (vWF) which are felt to be the major contributing factors to the increased risk of VTE [7, 16]. ABO blood grouping has been reported to influence activated protein C resistance [8], plasma lipid levels [11] and markers of inflammation including soluble intercellular adhesion molecule 1 (ICAM1), plasma soluble E-selection and P-selectin and tumour necrosis factor-alpha [11]. An additive effect on VTE risk and ABO blood group has also been described in association with factor V Leiden and prothrombin gene mutations [10, 17]. Finally, ABO blood grouping has more recently been reported to influence

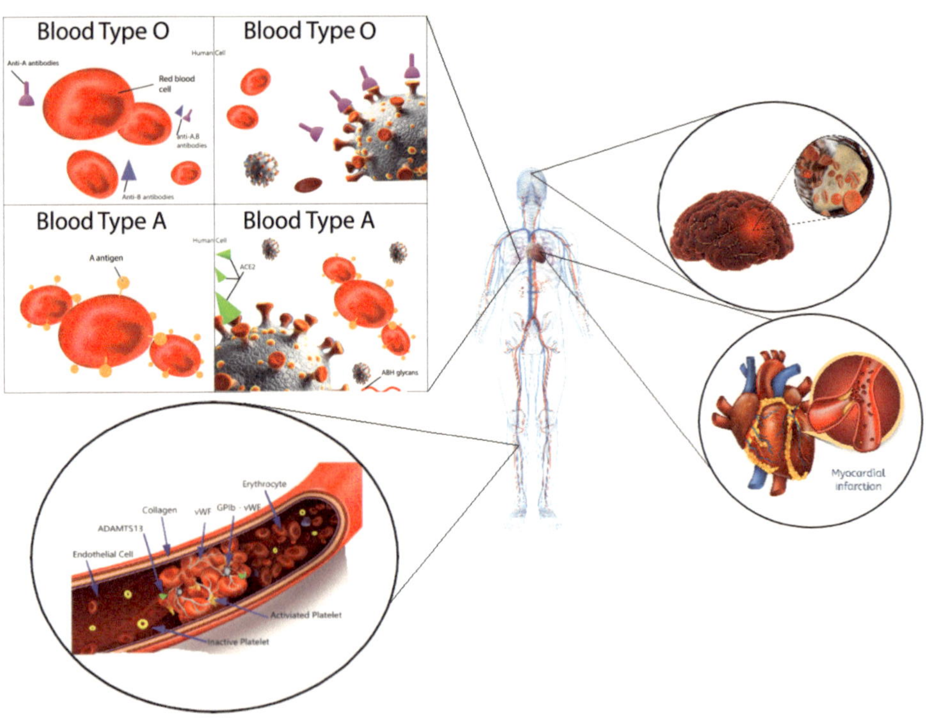

Figure 1.
ABO gene and their potential role in COVID-19, cerebral ischaemic disease, peripheral arterial disease, and myocardial infarction.

susceptibility to SARS-CoV-2 infection and an individual's propensity to more severe disease [18–20]. Proposed mechanisms include an additive risk of COVID-19 related thrombophilic complications in patients with a non-O blood group and a protective role of the O-blood group against the binding of the SARS-CoV-2 spike protein to lung epithelium angiotensin converting enzyme-2 receptor [18–20].

There are at least 35 minor blood group antigens in addition to ABO including Kidd, MNS, Duffy and Lewis [21] for which some, including Kidd and Lewis, have also been associated with CVD although the mechanisms of the associations are unclear [22–24].

This review will focus on the relationship between ABO blood types and CVD including coronary artery disease, ischaemic stroke, peripheral arterial disease and VTE (**Figure 1**). The contemporary relationship between SARS-CoV-2 infection, CVD and ABO blood grouping and the role of minor blood group antigens in the pathogenesis of CVD will also be discussed.

2. The ABO blood group system

The discovery of the ABO blood group system in 1900 by Austrian physician Karl Landsteiner saw him awarded of a Nobel Prize in physiology and medicine thirty years later [25]. The ABO blood group system consists of three main alleles A, B and O with codominant A and B alleles resulting in an inheritance pattern consisting of six genotypes and four major blood types [4, 25]. The ABO locus is found on chromosome 9 (9q34.1-q34.2) and codes for 2 glycosyltransferases A and B that transfer N-acetyl-D-galactosamine and D-galactose to a H antigen acceptor site on red blood cells (RBC) producing A and B surface antigens and blood group types, respectively (**Figure 2**) [26]. Lack of glycosyltransferase activity results in an unmodified ABO H antigen precursor and an O blood group type (O standing for 'Ohne', the German word for 'without') [25, 26]. Although ABO blood group antigens are RBC antigens, they are also expressed on human tissue including epithelial and endothelial cells [4].

The four basic ABO blood groups are O, A, B and AB. The ABO blood phenotypes have multiple subtypes, including A1, A2 and A3, in which 80% of blood group A cases are A1 in subtype, and O1, O2 and O3, in which O1 accounts for 95% of blood group O cases (**Table 1**) [4, 26]. From an evolutionary perspective, the oldest blood groups are A and O with A1 subtype considered the ancestral blood group [4, 7]. Single nucleotide polymorphisms (SNPs) of the ABO gene define the major haplotypes of European ancestry populations [7]. A substitution from proline to leucine at amino acid 156 results in a change from A1 to the less common A2 allele [5, 7]. Substitutions from glycine to serine at amino acid 235, leucine to methionine at amino acid 266 and glycine to alanine at amino acid 268 results in B allele subtypes [7, 27]. In contrast, the O1 type is a consequence of a frameshift deletion of guanine (cdel261G, p88fs118Stop) which translates to a protein without enzymatic activity [4, 5, 7].

There are significant geographic and racial variations in the distribution of blood groups across the world [4, 25]. Contributing factors include migration over the time of humankind's existence and processes of natural selection influenced by environmental factors such as climate and major diseases including malaria for which the O blood group confers protection [25]. Blood group A is predominant in Northern and Central Europe, B in Central Asia and O in Africa, South America, and Australia [4, 5, 25]. However, there are isolated populations within each continent that have a completely different blood group [4]. For example, the O blood group is common in different areas of Europe including Scandinavia and Switzerland [10, 25]. The most recent blood group, AB, appears to have arisen when A blood group populations in Europe migrated and mixed with B blood group populations of Asia [4].

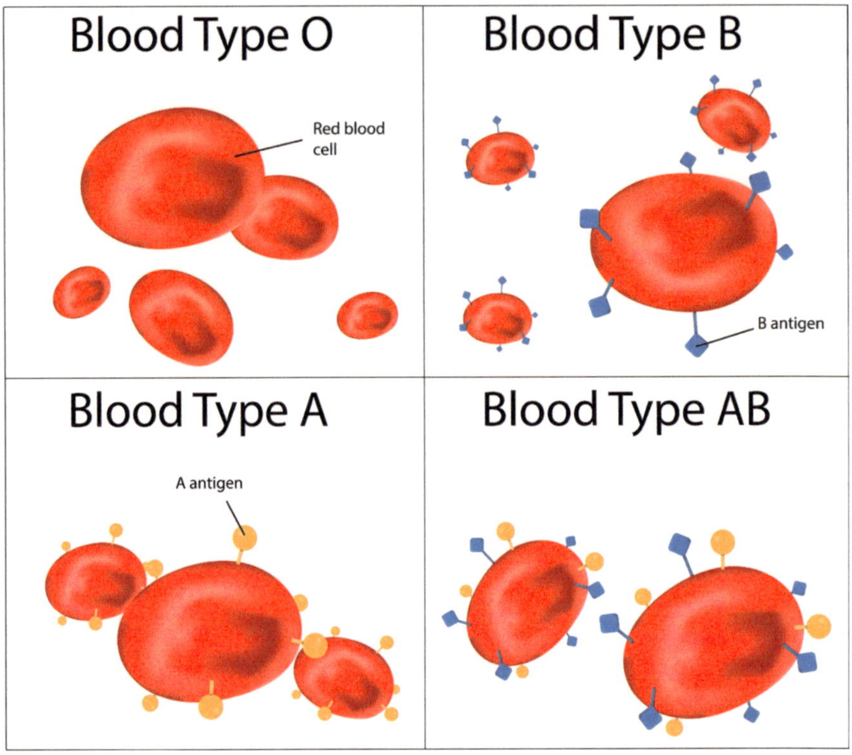

Figure 2.
Transfer of N-acetyl-D-galactosamine and D-galactose to a H antigen acceptor site on red blood cells (RBC) produces A and B surface antigens and blood group types, respectively. Lack of such a transfer result in an unmodified ABO H antigen precursor and the O blood group type.

ABO blood groups [25]	
A	A1, A2, A3 and other rare types including A4, A5, A6, Z, X, End, Boutu, g and i
B	B1, B2, B3 and rare types w, x, v and m
Other subtypes	O1, O2, O3 and other types including Yy, Hh, Xx and Bombay

Table 1.
The main ABO groups and their subtypes.

Due to the association between clotting and ABO blood group, these geographic and racial differences may contribute to ethnic differences seen in rates of CVD.

3. ABO blood groups and arterial diseases

There have been numerous retrospective and prospective studies and meta-analyses demonstrating an association between non-O blood groups and CVD events [5, 8–12, 24]. These studies have consistently demonstrated an association between non-O blood groups and an increased risk of arterial disease including myocardial infarction, coronary heart disease, peripheral arterial disease, and ischaemic stroke (**Table 2**). Possible mechanisms for an association with CVD include vWF-related thrombosis and modulation of platelet function through other platelet proteins which also express ABO antigens such as glycoprotein IIb [5]. The following two sections discuss the relationship between ABO blood groups and arterial diseases.

Study and year	Type of study	Subjects	Age range	Country	Findings (non-O versus O)
Medalie et al. 1971 [24]	5-year prospective observational	10,000 males	>40 y	Israel	↑ MI and angina pectoris in A1, B, A1Jk^{a-}, BJk^{a-}
Wu et al. 2007 [8]	Meta-analysis	45 arterial studies with 9720 subjects	Multiple ages	Multiple countries	↑ MI (OR 1.25), PAD (OR 1.45) & cerebral arterial ischaemia (OR1.14) in non-blood O
Wiggins et al. 2009 [30]	Case control study (10 y follow-up)	1063 MI, 469 ischaemic & 91 haemorrhagic stroke vs. 3462 C	Age range 30–79	United States	↑ MI in blood group A^{11} & ↑ ischaemic stroke in blood group B compared to O1O1
He et al. 2012 [11]	26-year prospective observational	62,073 female nurses	30–55 y at baseline	United States	↑ coronary heart disease in A, B and AB blood groups
He et al. 2012 [11]	20-year prospective observational	27,428 males	40–75 y at baseline	United States	↑ coronary heart disease in blood groups A, B & AB
Zakai et al. 2014 [12]	Case cohort study (median 5.8 y follow-up)	646 stroke & 989 non-stroke controls (REGARD study)	Mean ages 63.6–66.8 y	United States	↑ stroke risk in blood group AB versus O with hazard ratio greater in those without diabetes mellitus
Vasan et al. 2016 [10]	SCANDAT2 (12.6 y median follow-up)	1112072 blood donors	Mean age 33 y at baseline	Sweden and Denmark	↑ MI & stroke in non-O blood group
Capuzzo et al. 2016 [6]	> 2-year observational (median 5.3 y)	249 blood donors with Cardiorisk score > 20	35–55 y	Italy	↑ clinical or subclinical CVD events including ACS, cerebral ischaemia, cardiac arrhythmia & PAD in non-O group
Chen et al. 2016 [9]	Meta-analysis	17 studies of 225,810 participants	Multiple ranges	Multiple countries	↑ coronary heart disease in blood group A (OR 1.14) vs. O (OR 0.89)
Lin et al. 2017 [28]	Acute observational	1209 ST-elevation MI patients	Mean age 55 y	China	↑ post-MI spontaneous recanalization in O group
Fu et al. 2020 [29]	Case control study	61 HR & 600 randomly selected MI C	75.6 y (HR) v 66.2 y C	China	↑ HR following MI in blood group A

ACS = acute coronary syndrome; C = control subjects; CVD = cardiovascular disease; HR = heart rupture; MI = myocardial infarction; OR = Odds Ratio; PAD = peripheral arterial disease; SCANDAT = Scandinavian Donations and Transfusions; y = years.

Table 2.
Subject characteristics and non-O versus O blood group findings for the studies presented in the ABO blood groups and arterial diseases section (by year of publication).

4. ABO and coronary artery disease

In 1971, Medalie et al. reported the findings of a five-year prospective study of 10,000 Israeli male government employees aged ≥40 years who were born in six different regions (Eastern, Central and South-eastern Europe, Israel, Asia, and North Africa) [24]. The study found that subjects with blood groups A1, B, and A1B tended to have higher rates of myocardial infarction and those with A1 and B had higher rates of angina pectoris when compared to other blood groups [24]. Further, subjects who were negative for the Kidd glycoprotein (JK^{a-}), a red blood cell urea transporter, had the highest rates of myocardial infarction and angina pectoris and adding this group to the ABO system ($A1Jk^{a-}$, BJk^{a-} and, particularly, $A1BJk^{a-}$) was associated with very high incidence rates [24].

A pooled analysis of two large prospective United States (US) cohort studies, the Nurses' Health Study (NHS) which included 62,073 women and the Health Professionals Follow-up Study (HPFS) which included 27,428 men, both of which had >20 years follow-up, also found that the ABO blood group was significantly associated with an increased risk of coronary heart disease for men and women [11]. A limitation of these two patient cohorts was the self-reporting of ABO and Rh factor status. However, a validation analysis of a subsample of 98 subjects found a 93% serologically confirmed ABO consistency for NHS and 90% consistency for HPFS. The combined analysis found that those with blood group A, B or AB were more likely to develop coronary heart disease and this risk was independent of age, level of physical activity, alcohol or smoking consumption or diabetes history [5, 11]. A meta-analysis, performed by the same authors, of 7 cohort studies including NHS and HPFS which combined a total of 114,648 individuals and 5,741 coronary heart disease cases found a significant pooled relative risk for coronary heart disease in patients with a non-O blood group of 1.1 (95% CI 1.05–1.18, $p = 0.0001$). Subjects with an O blood group had a lower risk for coronary heart disease when compared to B or AB with a trend seen for blood group A [5, 11].

ABO blood group status may be clinically relevant in subjects with concurrent cardiovascular risk factors. A study of 289 Italian blood donors with a high cardiovascular risk score (≥20) found that those with a non-O blood group had an increased risk of CVD events (including acute coronary syndrome, cerebral ischaemia, cardiac arrhythmias and supraaortic trunk or iliac artery stenosis) during a median follow-up of 5.3 years [6]. ABO blood group status may also influence CVD-related patient outcomes. In a study of 1209 patients with acute myocardial infarction, Lin et al. found a higher rate of spontaneous recanalization following myocardial infarction in association with the O blood group whereas the rate of spontaneous recanalization was lower in subjects with an A blood group [28]. Blood group A type has also been associated with an increased risk of heart rupture following myocardial infarction [29]. Hence, ABO blood type may not only predispose susceptible individuals to an increased risk of CVD events but may also influence post-myocardial infarction outcomes.

5. ABO, stroke, and arterial disease in general

ABO blood group status has been shown to influence stroke risk. An association of AB blood group with stroke (adjusted Hazards Ratio [aHR] 1.83, 95% CI 1.01–3.30) was found in the (Reasons for Geographic And Racial Differences in Stroke [REGARDS]) study which involved 30,239 US participants followed up over 5.8 years [12]. This finding remained significant after adjustment for age, gender, race region and Framingham stroke risk factors (systolic blood pressure, taking antihypertensive medication, diabetes, current smoking, atrial fibrillation and left ventricular hypertrophy). The association was greater in those participants without diabetes mellitus (aHR 3.33 95% CI, 1.61–6.88). Factor VIII levels accounted for 60% of the AB associated stroke risk [12]. Another study by Wiggins et al. identified an increased risk of ischaemic stroke in subjects with a B blood group (OR 1.59, 95% CI 1.17–2.17) [30].

Meta-analysis studies have also demonstrated associations between non-O blood groups and arterial diseases in general. Chen et al. performed a meta-analysis of 17 studies involving 225,810 participants and found that blood group A was associated with an increased risk of coronary artery disease (OR 1.14, 95% CI 1.03–1.26, $p = 0.01$) and blood group O a lower risk (OR 0.85, 95% CI 0.78–0.94, $p < 0.001$) [9]. These results remained significant after cases of myocardial infarction were excluded. Wu et al. conducted a systematic review and meta-analysis of 59 studies, both retrospective and prospective, reporting associations of ABO blood groups and arterial disease [8]. They found significant ORs of 1.25 (95% CI 1.14–1.36) for myocardial infarction ($n = 22$ studies), 1.45 (95% CI 1.35–1.46) for peripheral arterial disease ($n = 8$ studies) and 1.14 (95% CI 1.01–1.27) for cerebral ischaemia of arterial origin ($n = 7$ studies) in subjects with a non-O blood group [8].

6. ABO blood groups and venous thromboembolism

The relationship between the ABO blood groups and VTE is most probably stronger than that seen in arterial diseases [5]. The association between blood group subtypes and VTE has been well documented in a few genome-wide association (GWAS) [5, 31, 32], meta-analyses and/or case–control studies [7, 8, 10, 13] (**Table 3**). A French GWAS study involving 419 patients with early age onset of first deep vein thrombosis (DVT) who were compared to 1228 controls found that participants with blood type O had a 67% lower risk of VTE compared to n those with a non-O blood group [5, 31]. Relative to other non-O groups, subjects with the uncommon A2 subtype had a 47% lower risk [5, 31]. In the same study, Factor V Leiden mutations were also associated with increased risk of VTE although the authors did not investigate the potential for an additive risk in those having both a non-O blood group and a Factor V Leiden mutation [31]. A similar association of non-O blood group with VTE was found in another GWAS study comparing 1503 VTE subjects to 1459 age and gender matched controls [32]. In this study, the population attributable risk for VTE was highest for the non-O blood group, followed by blood type A, Factor V Leiden, and prothrombin G20210A [32].

There is evidence that the A2 subtype of blood group A is associated with a lower VTE risk when compared to the A1 allele [5, 33]. The A2 allele possesses a single base deletion near its carboxyl terminal (1061delC) which results in 30 to 50-fold less A-transferase activity than its A1 counterpart which suggests a correlation between the degree of H antigen glycosylation and VTE risk [33]. Data from 2 population-based case control studies that included 504 post-menopausal women with non-fatal VTE found that the B and AB blood groups were both associated with an increased risk of VTE (OR 1.82, 95% CI 1.29–2.57, and OR 2.7, 95% CI 1.73–4.21, respectively) when compared to O1O1 subjects [30]. Participants with A11, a subtype of the A1 allele, also carried a 79% increased risk of VTE (OR 1.79, 95% CI 1.41–2.26) [30]. An increased VTE risk was also identified in a population-based case control study of venous thrombosis (Leiden Thrombophilia Study) [34]. In this study, an increased thrombotic risk was found for non-O blood groups apart from those with genotypes homozygous to A2 or possession of any A2/O combination. Subjects with A1B/A2B and A1A1/A1A2 blood group genotypes had a 90–110%

Study and year	Type of study	Subjects	Age range	Country	Findings (non-O versus O)
Morelli et al. 2005 [34]	Case control study (LETS study)	471 patients & 471 C	Not reported	Nether-lands	↑ VTE for all non-O blood group except A2A2 or any A2O combination
Wu et al. 2007 [8]	Meta-analysis	21 VTE studies with 6720 subjects	Multiple	Multiple	↑ VTE in A1A2/A1B/BB (OR 2.44) & A!O/BO/A2B (OR 2.11) blood groups
Lima et al. 2009 [17]	Case control study	65 VTE patients & 51 C	Mean age 34 y (range 6–67 y)	Brazil	↑ VTE for FVL (OR 10.1) and double ↑ risk if also AB (OR 22.3)
Trégouët et al. 2009 [31]	Case control study (GWAS screening)	419 VTE patients & 1228 healthy C	Age at onset<50 y	France	↑ VTE in non-O blood group, relative lower risk VTE of A2 amongst non-O subjects and additive risk of FVL mutation
Wiggins et al. 2009 [30]	Case control study (10 y follow-up)	Peri/post-MP women 504 VTE & 2172 C	Age range 30–89 y	United States	↑ VTE in B (OR 1.82), AB (OR 2.7) & A11 (OR 1.79) compared to O1O1 blood group
Heit et al. 2012 [32]	Case control study (GWAS analysis)	1503 VTE & 1459 age & gender matched C	Mean age for VTE & C 55 y	United States	↑ VTE population attributable risk for non-O blood group followed by A blood group, FVL & prothrombin G20210A
Vasan et al. 2016 [10]	SCANDAT2 (12.6 y median follow-up)	1112072 blood donors	Mean age 33 y at baseline	Sweden and Denmark	↑ VTE including pregnancy-related VTE & DVT in non-O blood group
Sun et al. 2017 [13]	Retrospective observational (7 y)	1412 VTE & 199,248 C	Mean 57.3 y VTE & 47.5 C	China	↑ VTE in non-O blood group (OR 1.35)
Goumidi et al. 2021 [7]	Pooled analysis of 6 studies	5425 VTE & 8445 C	Not reported	Multiple countries	↑ VTE for A2 (OR 1.2), A1 & B (OR 1.8) & ↓ VTE for O2 (OR 0.8) compared to O1

C = control subjects; DVT = deep vein thrombosis; FVL = Factor V Leiden; GWAS = genome-wide association study; LETS = Leiden Thrombophilia Study; MP = menopausal; OR = Odds Ratio; PTE = pulmonary thromboembolism; SCANDAT = Scandinavian Donations and Transfusions; VTE = venous thromboembolism; y = years.

Table 3.
A summary of subject characteristics and non-O versus O blood group findings for the studies presented in the ABO blood groups and VTE section (by year of publication).

increased risk and those with BB/BO1/B02 genotypes had a 60% increased risk when compared to the OO genotype [34].

In a pooled analysis of 6 case–control/prospective studies of VTE, the A2 subgroup was found to be associated with a modest increase in VTE risk (OR 1.2) whereas the A1 and B subgroups had a 1.8-fold increased risk compared to the O1 subgroup. In contrast, O2 had a relative protective effect (OR 0.8) [7]. Both the A1 and B subgroups were associated with increased vWF and factor VIII plasma levels whereas only the A1 subgroup was associated with increased ICAM levels [7]. A meta-analysis of 21 VTE studies, 18 of which were retrospective in design, found a pooled OR of 1.79 (95% CI 1.56–2.05) for non-O blood group and VTE [7]. The combination of A1A1/A1B/BB genotypes had an OR of 2.44 (95% CI 1.79–3.33) and A1O/BO/A2B an OR of 2.11 (95% CI 1.66–2.68) for VTE [8].

The presence of a non-O blood group has been found to correlate with unprovoked PTE, provoked including pregnancy induced and recurrent VTE. In a Scandinavian 25-year follow-up study of >1.6 million blood donors, those with a non-O blood group had higher rates of pregnancy-related VTE events, DVT and pulmonary embolism [10]. The risks of recurrent pulmonary embolism and DVT provoked by comorbid illness were also higher in subjects with a non-O blood group [10]. A large hospital-based retrospective study of 200,660 Han Chinese patients including 1412 VTE subjects (600 with DVT, 441 with pulmonary embolism and 371 with both conditions) conducted between 2010 and 2016 found a significant association of non-O blood group with VTE (OR 1.35, 95% CI 1.21–1.54) [13]. Interestingly, subgroup analysis found a relatively greater non-O blood group risk for those with an unprovoked VTE (OR 1.86) compared to provoked VTE (OR 1.22). The OR for having a non-O blood group also appeared to decrease with age [13]. Finally, a Brazilian case control study comparing 65 subjects with a history of DVT to 51 controls showed a significant increased risk of VTE in the presence of Factor V Leiden mutation (OR 10.1) which doubled in those in whom the AB allele of the ABO blood group was also present (OR 22.3) [17].

Future research into the role of developing risk stratification models or algorithms, for example, by combining ABO and other genetic variables with patient comorbidity and arterial risk factors to identify individuals at higher risk of VTE is warranted. This may translate to improved cardiovascular disease management including decision making for VTE prophylaxis at the hospital and outpatient setting.

7. Von Willebrand factor, factor VIII and other factors associated with ABO blood group and CVD

The majority of studies that have been presented in this review implicate an increase in plasma levels of vWF and factor VIII by non-O blood group types as the likely mechanism for an increased risk of thromboembolic events [5–13]. VWF/factor VIII are important in the acute phase response to vessel injury [8]. VWF is a carrier of factor VIII, protects it from inactivation, can also recruit platelets to a site of vessel injury to induce a coagulation cascade responsible for clot formation [5]. VWF can also bind to the platelet receptor glycoprotein Iba, to form a bridge between platelets and the endothelium and participate in a thrombo-inflammatory response (**Figure 3**) [35, 36].

ABO blood groups may have a direct functional effect on circulating vWF and thereby modulate both vWF and factor VIII levels [5]. The mechanism by which the presence of an N-acetyl-D-galactosamine or D-galactose residue on glycans expressed on the H antigen acceptor site of a RBC influences plasma levels and/or bioactivity of vWF is unclear [16]. A plausible mechanism is that ABO glycans on vWF itself influence its release into plasma and/or its clearance [16]. VWF is derived from a pre-pro-polypeptide (ppvWF) synthesized in endothelial cells and megakaryocytes [16]. The expression of blood antigen A on vWF correlates with decreased ppvWF to vWF ratios as well as longer half-life and increased plasma levels of vWF [16].

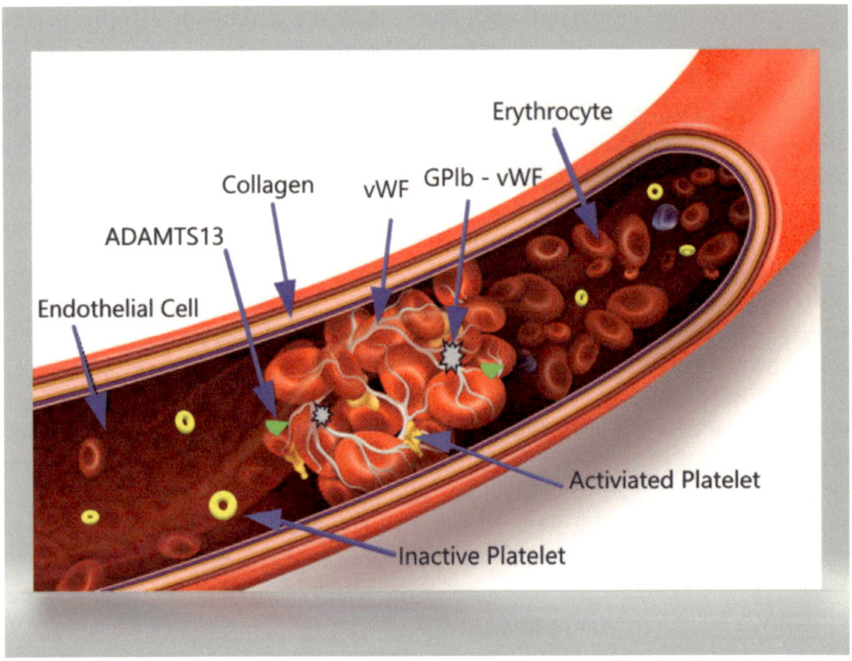

Figure 3.
*Von Willebrand factor (vWF) can bind to the platelet receptor glycoprotein Ib (GPIb) to form a bridge between platelets and the endothelium and a thrombo-inflammatory response. The protease ADAMTTS13 (**a d**isintegrin **a**nd **m**etalloproteinase with a **t**hrombo**s**pondin type 1 motif, member **13**) is responsible for the proteolysis and clearance of vWF from the circulation.*

It has been postulated that the presence of A and B terminal carbohydrate antigens influence the proteolysis of vWF by its major protease ADAMTTS13 (**a d**isintegrin **a**nd **m**etalloproteinase with a **t**hrombo**s**pondin type 1 motif, member **13** - also known as von Willebrand factor-cleaving protease [VWFCP]) [5]. Individuals lacking glycosyltransferase activity (O blood group) have higher levels of ADAMTS13 activity suggesting that ABO glycosyltransferase activity can indirectly modify, for example, inhibit proteolytic activity and reduce vWF clearance from the circulation [5]. However, there are studies which have not supported this as the underlying mechanism [37]. Future research into other, yet to be determined, ABO blood group and VWF-related associations may unravel the underlying mechanism of a non-O blood group increased risk of thrombosis.

Blood group antigens are also associated with elevated plasma levels of markers of inflammation including endothelial cell and platelet-derived adhesion molecules [5]. Elevated plasma levels of adhesion molecules including soluble P-selectin, soluble ICAM-1 and tumour necrosis factor-alpha are associated with ABO genotype, which may result in arterial and venous thrombosis and an increased CVD risk [5, 11]. However, in studies reporting the circulating expression levels of sICAM-1 and soluble P-selectin, the blood group A has paradoxically been found to be associated with lower circulating expression levels of sICAM-1 and soluble P-selectin when compared to the O blood group [38, 39]. This contradictory finding has been described in healthy Chinese populations and a study of Caucasian women without a history of chronic disease [38, 39]. A postulated explanation for why blood group A, in general, may be associated with lower circulating inflammatory markers despite carrying a higher CVD risk is that higher levels of sICAM-1 or soluble P-selectin expression are limited to those with significant CVD risk factors and/or symptomatic CVD. The concentrations of membrane forms of these adhesion molecules may also higher and thereby still mediate leucocyte migration and adhesion at an endothelial level [39]. It is also possible that reduced levels of sICAM-1 increase the adhesion of leukocytes on endothelial surfaces which may result in increased arterial inflammation [40].

Platelet glycoproteins including GPIIb, and platelet endothelial cell adhesion molecule are known to carry ABO blood group antigens and may be involved in thrombosis through modulation of the GPIIb-GPIIIa fibrinogen receptor complex [5]. A relationship between ABO blood group and angiotensin converting enzyme activity has also been reported which implicates a role for ABO blood group in the regulation of arterial risk factors such as hypertension [41].

Epidemiological studies have demonstrated an association between an elevated serum cholesterol including low-density lipoprotein cholesterol and non-O blood groups [5, 42]. These findings implicate ABO genotypes in the modulation of plasma lipids. ABO blood groups are also associated with phytosterol levels which have also been reported to modify cardiovascular risk [5].

8. ABO blood group, COVID-19, and CVD

There is growing evidence that ABO blood groups may play a role in the susceptibility to and severity of SARS-CoV-2 infection [18–20]. Individuals with blood group O have a lower risk and those with blood group A carry a higher risk of SARS-CoV-2 infection [18]. A systematic review by the International Society of Blood Transfusion (ISBT) COVID-19 working group recently reported that subjects with blood group A had a higher rate of SARS-CoV-2 infection as well as an increased risk of requiring mechanical ventilation, continuous renal replacement therapy and prolonged intensive care unit stay [18]. Postulated mechanisms include an increase in angiotensin converting enzyme-1 levels in blood group A patients and an increased risk of cardiovascular, thromboembolic, and inflammatory complications [18]. It is plausible that possession of a non-O blood group and associated increase in vWF and factor VIII levels have an additive effect to the thrombophilia caused by SARS-CoV-2 and results in an increased risk of COVID-related CVD complications [18–20]. A recent hypothesis for an association between non-O blood group type and risk of SARS-CoV-2 infection is that anti-A and/or anti-B antibodies, which are present in patients with blood group O bind to a corresponding antigen, for example the angiotensin-converting-enzyme-2-receptor, on the SARS-CoV-2 viral envelope which then inhibits viral entry into lung epithelium (**Figure 4**) [18]. Although ABH antigen structures are yet to be described on the SARS-CoV-2 protein, the spike protein has been reported to possess N-glycans and N-glycosylation sites which could potentially interact with anti-A and anti-B antigens and thus confer protection against infection in individuals with the blood group O type [18]. The possibility that blood group A patients also have higher rates of underlying comorbidities which significantly contribute to COVID-related complications cannot be excluded [18].

A single-centre study from Bangladesh which evaluated 438 patients with SARS-CoV-2 infection also found a significantly higher rate of blood group A amongst COVID-19 patients compared to the general population [20]. However, ABO blood groups were not associated with type of presentation or recovery from infection [20]. Conversely, an observational study of 14,112 individuals tested for SARS-CoV-2 in the New York Presbyterian hospital system found that risk of intubation was increased for AB and B blood groups but decreased for A when compared to O blood group and risk of death was increased for those with AB blood group and decreased for A and B blood groups [19]. Interestingly, Rhesus status, which is not implicated

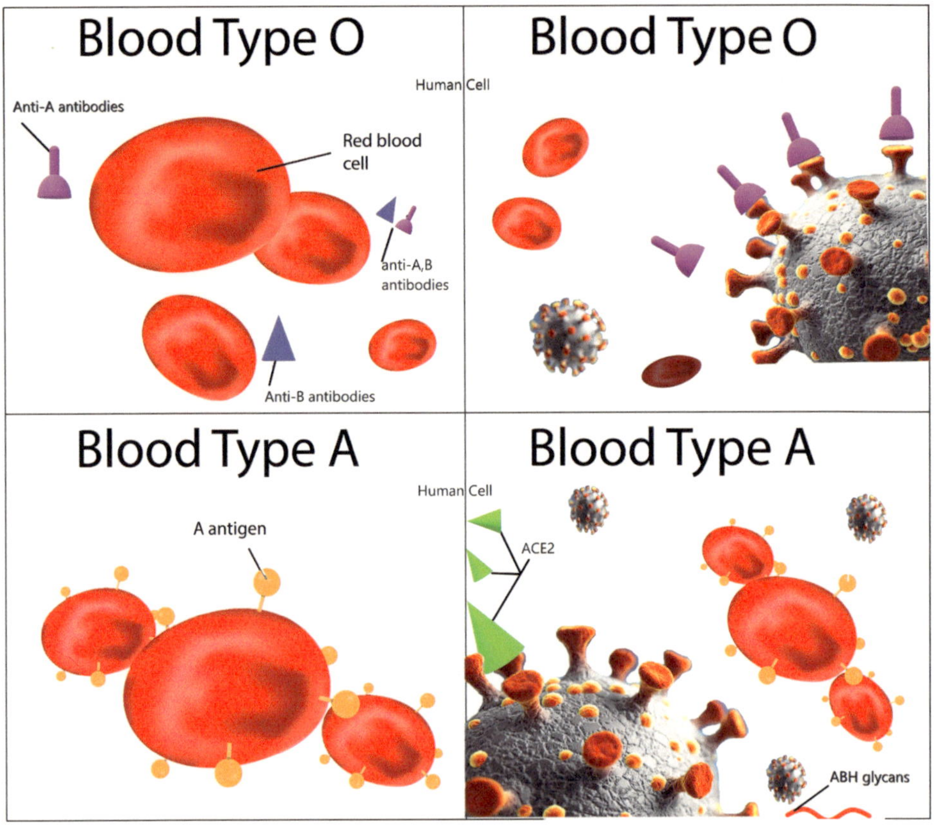

Figure 4.
A hypothesis for an association between non-O blood group type and risk of SARS-CoV-2 infection is that anti-A and/or anti-B antibodies may bind to the angiotensin-converting-enzyme-2-receptor on the SARS-CoV-2 viral envelope and thereby inhibit viral entry into lung epithelium. Top left shows blood group O with anti-A and anti-B antibodies in the plasma. Top right illustrates anti-A antibodies of blood group O which potentially competitively bind to the spike protein of SARS-CoV-2 and thus inhibit infection. Lower left shows blood group A with A-antigens on the membranes of red blood cells. Lower right depicts ABH glycans on the SARS-CoV-2 spike protein which potentially competitively bind to angiotensin converting enzyme 2 (ACE2) receptors.

in CVD risk, correlated with COVID-19 risk [19]. Rhesus-negative subjects had a lower risk of SARS-CoV-2 infection, intubation, and death [19]. The mechanism of the relationships of ABO blood group, Rhesus status and SARS-CoV-2 infection is unknown. Further research into the relationship between blood groups and risk of SARS-CoV-2 infection and COVID-19 related complications is warranted.

9. Minor blood group antigens and CVD

There are over 35 minor blood group antigens on red blood cells [21], some of which including P and Lewis, are widely distributed in other human cells and body fluids [43, 44]. Minor blood groups have been associated with several diseases ranging from malignancy to peptic ulcer disease, infection and CVD [43, 44]. As discussed previously, subjects with a combination of blood groups A or B and the Kidd antigen Jk^{a-} status have been shown to be at increased risk of myocardial infarction [24]. Sialyl-Le^x (sLe^x), an antigen of the Lewis blood group system, is a major ligand for the cellular adhesion molecules E, P and L-selectin which are involved in the adhesion of leucocytes to endothelium [44]. The Duffy blood group glycoprotein is a chemokine receptor on RBCs that is involved in the recruitment of leucocytes to sites of inflammation [44]. Although the exact mechanisms are unclear, these biological characteristics offer explanation why the Lewis and Duffy blood group antigens may be associated with an increased risk of CVD.

The Lewis blood group system, which was first discovered by Mourant in 1946, is classified into four phenotypes [Le(a-b-), Le(a + b-), Le(a-b+), Le(a + b+)] determined by two genetic systems closely related on the short arm of chromosome 19 [45]. In a study of 3385 Danish males, the Le(a-b-) phenotype was associated with an increased risk of mortality from ischaemic heart disease [46]. In another Danish study involving 702 participants (72% male), the Le(a-b-) was associated with self-reported non-fatal stroke [47]. However, a North Indian cross-sectional study that compared 161 patients with angiographically-proven coronary artery disease with 71 control subjects with normal angiography, found that the lack of Lewis antigen expression was associated with coronary artery disease in female subjects only [48]. A lack of Lewis antigen expression has also been associated with a higher body mass index, weight gain over time, a lower level of physical activity, type 2 diabetes mellitus and hypertriglyceridemia all of which associated with an increased risk of CVD [49–52].

The Duffy antigen, located on the long arm of chromosome 1 [53], has also been implicated in CVD risk [54]. There are four Duffy phenotypes [Fy(a-b-), Fy(a + b-), Fy(a-b+), Fy(a + b+) [53]. A study of 5301 African American participants found that Duffy negative subjects with a neutrophil: lymphocyte ratio ≥ 1.77 were more likely to have coronary artery disease and stroke [54]. Lack of Duffy expression has also been associated with chronic organ damage, in particular renal dysfunction, in subjects with sickle-cell disease [55].

To the best of our knowledge, there have been no studies to date addressing the relationship of minor blood groups and COVID-19.

The limitations of studies evaluating the role of minor blood groups in CVD include their observational cohort design, case selection, outcome definitions, cohort sizes and influence of population attributable factors.

10. Conclusion

Subjects with non-O blood group type have an increased risk of arterial and venous thromboembolism. Blood groups A1, B and AB are at particularly increased risk of CVD events whereas blood group O confers a protective effect. Postulated mechanisms of underlying the relationship between ABO blood group and CVD include vWF and factor VIII activity and elevations in circulating inflammatory markers and plasma lipids. Comorbidities including arterial risk factors and predisposing factors to VTE such as concomitant Factor V Leiden mutations may have an additive effect to thromboembolic risk. Minor blood group types including Kidd, Lewis and Duffy are also been associated with CVD. The relationship of non-O blood group type and SARS-CoV-2 infection warrants further research. Future directions include the development and implementation of risk stratification algorithms for thromboembolic risk such as ABO blood group and other factors associated with arterial or venous disease in a hospital or outpatient setting.

Author details

Dennis J. Cordato[1,2,3], Wissam Soubra[3,4,5], Sameer Saleem[1,2,3] and Kaneez Fatima Shad[3,5,6]*

1 Department of Neurophysiology, Liverpool Hospital, NSW, Australia

2 South Western Sydney Clinic School, University of New South Wales, Sydney, NSW, Australia

3 Ingham Institute for Applied Medical Research, Liverpool, NSW, Australia

4 A Healthy Step Clinic, NSW, Australia

5 School of Life Sciences, University of Technology Sydney, NSW, Australia

6 School of Behavioural and Health Sciences, Faculty of Health Sciences, Australian Catholic University, NSW, Australia

*Address all correspondence to: kaneez.fatimashad@uts.edu.au; ftmshad@gmail.com

References

[1] Roth GA, Mensah GA, Johnson CO, Addolorato G, Ammirati E, Baddour LM, Barengo NC, Beaton AZ, Benjamin EJ, Benziger CP, Bonny A, Brauer M, Brodmann M, Cahill TJ, Carapetis J, Catapano AL, Chugh SS, Cooper LT, Coresh J, Criqui M, DeCleene N, Eagle KA, Emmons-Bell S, Feigin VL, Fernández-Solà J, Fowkes G, Gakidou E, Grundy SM, He FJ, Howard G, Hu F, Inker L, Karthikeyan G, Kassebaum N, Koroshetz W, Lavie C, Lloyd-Jones D, Lu HS, Mirijello A, Temesgen AM, Mokdad A, Moran AE, Muntner P, Narula J, Neal B, Ntsekhe M, Moraes de Oliveira G, Otto C, Owolabi M, Pratt M, Rajagopalan S, Reitsma M, Ribeiro ALP, Rigotti N, Rodgers A, Sable C, Shakil S, Sliwa-Hahnle K, Stark B, Sundström J, Timpel P, Tleyjeh IM, Valgimigli M, Vos T, Whelton PK, Yacoub M, Zuhlke L, Murray C, Fuster V; GBD-NHLBI-JACC Global Burden of Cardiovascular Diseases Writing Group. Global burden of cardiovascular diseases and risk factors, 1990-2019: update from the GBD 2019 Study. J Am Coll Cardiol 2020; 76(25):2982-3021

[2] European Association for Cardiovascular Prevention & Rehabilitation, Reiner Z, Catapano AL, De Backer G, Graham I, Taskinen MR, Wiklund O, Agewall S, Alegria E, Chapman MJ, Durrington P, Erdine S, Halcox J, Hobbs R, Kjekshus J, Filardi PP, Riccardi G, Storey RF, Wood D; ESC Committee for Practice Guidelines (CPG) 2008-2010 and 2010-2012 Committees. ESC/EAS Guidelines for the management of dyslipidaemias: the Task Force for the management of dyslipidaemias of the European Society of Cardiology (ESC) and the European Atherosclerosis Society (EAS). Eur Heart J 2011; 32(14):1769-818.

[3] Inouye M, Abraham G, Nelson CP, Wood AM, Sweeting MJ, Dudbridge F, Lai FY, Kaptoge S, Brozynska M, Wang T, Ye S, Webb TR, Rutter MK, Tzoulaki I, Patel RS, Loos RJF, Keavney B, Hemingway H, Thompson J, Watkins H, Deloukas P, Di Angelantonio E, Butterworth AS, Danesh J, Samani NJ; UK Biobank CardioMetabolic Consortium CHD Working Group. Genomic risk prediction of coronary artery disease in 480,000 adults: implications for primary prevention. J Am Coll Cardiol 2018; 72(16):1883-1893.

[4] Dean L. Blood Groups and Red Cell Antigens [Internet]. Bethesda (MD): National Center for Biotechnology Information (US); 2005. Chapter 5. The ABO blood group. Available from: https://www.ncbi.nlm.nih.gov/books/NBK2267/

[5] Zhang H, Mooney CJ, Reilly MP. ABO blood groups and cardiovascular diseases. Int J Vasc Med 2012; 2012:641917.

[6] Capuzzo E, Bonfanti C, Frattini F, Montorsi P, Turdo R, Previdi MG, Turrini E, Franchini M. The relationship between ABO blood group and cardiovascular disease: results from the Cardiorisk program. Ann Transl Med 2016; 4(10):189.

[7] Goumidi L, Thibord F, Wiggins KL, Li-Gao R, Brown MR, van Hylckama Vlieg A, Souto JC, Soria JM, Ibrahim-Kosta M, Saut N, Daian D, Olaso R, Amouyel P, Debette S, Boland A, Bailly P, Morrison AC, Mook-Kanamori DO, Deleuze JF, Johnson A, de Vries PS, Sabater-Lleal M, Chiaroni J, Smith NL, Rosendaal FR, Chasman DI, Trégouët DA, Morange PE. Association between ABO haplotypes and the risk of venous thrombosis: impact on disease risk estimation. Blood 2021; 137(17):2394-2402.

[8] Wu O, Bayoumi N, Vickers MA, Clark P. ABO(H) blood groups and

vascular disease: a systematic review and meta-analysis. J Thromb Haemost 2008; 6(1):62-69.

[9] Chen Z, Yang SH, Xu H, Li JJ. ABO blood group system and the coronary artery disease: an updated systematic review and meta-analysis. Sci Rep 2016; 6:23250.

[10] Vasan SK, Rostgaard K, Majeed A, Ullum H, Titlestad KE, Pedersen OB, Erikstrup C, Nielsen KR, Melbye M, Nyrén O, Hjalgrim H, Edgren G. ABO blood group and risk of thromboembolic and arterial disease: a study of 1.5 million blood donors. Circulation 2016; 133(15):1449-1457.

[11] He M, Wolpin B, Rexrode K, Manson JE, Rimm E, Hu FB, Qi L. ABO blood group and risk of coronary heart disease in two prospective cohort studies. Arterioscler Thromb Vasc Biol 2012; 32(9):2314-2320.

[12] Zakai NA, Judd SE, Alexander K, McClure LA, Kissela BM, Howard G, Cushman M. ABO blood type and stroke risk: the REasons for Geographic And Racial Differences in Stroke Study. J Thromb Haemost 2014; 12(4):564-570.

[13] Sun X, Feng J, Wu W, Peng M, Shi J. ABO blood types associated with the risk of venous thromboembolism in Han Chinese people: A hospital-based study of 200,000 patients. Sci Rep 2017; 7:42925.

[14] Sun Y, Mao P, Lu J, Dai J, Teng H, Jiang Q. ABO blood type and ABO gene with susceptibility to deep vein thrombosis following orthopedic surgery: a case-control study in Chinese Han population. Sci China Life Sci 2015; 58(4):390-391.

[15] Samama MM; for the Sirius Study Group. An epidemiologic study of risk factors for deep vein thrombosis in medical outpatients: the Sirius study. Arch Intern Med 2000; 160(22):3415-3420.

[16] Stowell CP and Stowell SR. Biologic roles of the ABH and Lewis histo-blood group antigens Part I: infection and immunity. Vox Sang 2019; 114(5):426-442.

[17] Lima MB, de Oliveira-Filho AB, Campos JF, Melo FC, Neves WB, Melo RA, Lemos JA. Increased risk of venous thrombosis by AB alleles of the ABO blood group and Factor V Leiden in a Brazilian population. Genet Mol Biol 2009; 32(2):264-267.

[18] Goel R, Bloch EM, Pirenne F, Al-Riyami AZ, Crowe E, Dau L, Land K, Townsend M, Jecko T, Rahimi-Levene N, Patidar G, Josephson CD, Arora S, Vermeulen M, Vrielink H, Montemayor C, Oreh A, Hindawi S, van den Berg K, Serrano K, So-Osman C, Wood E, Devine DV, Spitalnik SL; ISBT COVID-19 Working Group. ABO blood group and COVID-19: a review on behalf of the ISBT COVID-19 working group. Vox Sang 2021 Feb 12:10.1111/vox.13076. doi: 10.1111/vox.13076. Epub ahead of print.

[19] Zietz M, Zucker J, Tatonetti NP. Associations between blood type and COVID-19 infection, intubation, and death. Nat Commun 2020; 11(1):5761.

[20] Mahmud R, Rassel MA, Monayem FB, Sayeed SKJB, Islam MS, Islam MM, Yusuf MA, Rahman S, Islam KMN, Mahmud I, Hossain MZ, Chowdhury AH, Kabir AKMH, Ahmed KGU, Rahman MM. Association of ABO blood groups with presentation and outcomes of confirmed SARS CoV-2 infection: A prospective study in the largest COVID-19 dedicated hospital in Bangladesh. PLoS One 2021; 16(4):e0249252.

[21] Holt SG, Kotagiri P, Hogan C, Hughes P, Masterson R. The potential role of antibodies against minor blood

group antigens in renal transplantation. Transpl Int 2020; 33(8):841-848.

[22] Ellison RC, Zhang Y, Myers RH, Swanson JL, Higgins M, Eckfeldt J. Lewis blood group phenotype as an independent risk factor for coronary heart disease (the NHLBI Family Heart Study). Am J Cardiol 1999; 83(3):345-348.

[23] Enevold C, Nielsen CH, Molbo D, Lund R, Bendtzen K, Fiehn N-, Holmstrup P. Lewis and AB0 blood group-phenotypes in periodontitis, cardiovascular disease, obesity and stroke. Sci Rep. 2019; 9(1):6283.

[24] Medalie JH, Levene C, Papier C, Goldbourt U, Dreyfuss F, Oron D, Neufeld H, Riss E. Blood groups, myocardial infarction and angina pectoris among 10,000 adult males. N Engl J Med 1971; 285(24):1348-1353.

[25] Farhud DD, Zarif Yeganeh M. A brief history of human blood groups. Iran J Public Health 2013; 42(1):1-6.

[26] Ségurel L, Thompson EE, Flutre T, Lovstad J, Venkat A, Margulis SW, Moyse J, Ross S, Gamble K, Sella G, Ober C, Przeworski M. The ABO blood group is a trans-species polymorphism in primates. Proc Natl Acad Sci U S A 2012; 109(45):18493-18498.

[27] Yamamoto F. Molecular genetics of the ABO histo-blood group system. Vox Sang 1995; 69(1):1-7.

[28] Lin XL, Zhou BY, Li S, Li XL, Luo ZR, Li JJ. Correlation of ABO blood groups with spontaneous recanalization in acute myocardial infarction. Scand Cardiovasc J 2017; 51(4):217-220.

[29] Fu Y, Chen M, Sun H, Guo Z, Gao Y, Yang X, Li K, Wang L. Blood group A: a risk factor for heart rupture after acute myocardial infarction. BMC Cardiovasc Disord 2020; 20(1):471.

[30] Wiggins KL, Smith NL, Glazer NL, Rosendaal FR, Heckbert SR, Psaty BM, Rice KM, Lumley T. ABO genotype and risk of thrombotic events and hemorrhagic stroke. J Thromb Haemost 2009; 7(2):263-269.

[31] Trégouët DA, Heath S, Saut N, Biron-Andreani C, Schved JF, Pernod G, Galan P, Drouet L, Zelenika D, Juhan-Vague I, Alessi MC, Tiret L, Lathrop M, Emmerich J, Morange PE. Common susceptibility alleles are unlikely to contribute as strongly as the FV and ABO loci to VTE risk: results from a GWAS approach. Blood 2009; 113(21):5298-5303.

[32] Heit JA, Armasu SM, Asmann YW, Cunningham JM, Matsumoto ME, Petterson TM, De Andrade M. A genome-wide association study of venous thromboembolism identifies risk variants in chromosomes 1q24.2 and 9q. J Thromb Haemost 2012; 10(8):1521-1531.

[33] Yamamoto F, McNeill PD, Hakomori S. Human histo-blood group A2 transferase coded by A2 allele, one of the A subtypes, is characterized by a single base deletion in the coding sequence, which results in an additional domain at the carboxyl terminal. Biochem Biophys Res Commun 1992; 187(1):366-374.

[34] Morelli VM, De Visser MC, Vos HL, Bertina RM, Rosendaal FR. ABO blood group genotypes and the risk of venous thrombosis: effect of factor V Leiden. J Thromb Haemost 2005; 3(1):183-185.

[35] Buchtele N, Schwameis M, Gilbert JC, Schörgenhofer C, Jilma B. Targeting von Willebrand Factor in ischaemic stroke: focus on clinical evidence. Thromb Haemost 2018; 118(6):959-978.

[36] Denorme F, Vanhoorelbeke K, De Meyer SF. von Willebrand Factor and platelet glycoprotein Ib: A

thromboinflammatory axis in stroke. Front Immunol 2019; 10:2884.

[37] Badirou I, Kurdi M, Rayes J, Legendre P, Christophe OD, Lenting PJ, Denis CV. von Willebrand factor clearance does not involve proteolysis by ADAMTS-13. J Thromb Haemost 2010; 8(10):2338-2340.

[38] Zhang W, Xu Q, Zhuang Y, Chen Y. Novel association of soluble intercellular adhesion molecule 1 and soluble P-selectin with the ABO blood group in a Chinese population. Exp Ther Med 2016; (2):909-914.

[39] Paré G, Chasman DI, Kellogg M, Zee RY, Rifai N, Badola S, Miletich JP, Ridker PM. Novel association of ABO histo-blood group antigen with soluble ICAM-1: results of a genome-wide association study of 6,578 women. PLoS Genet. 2008; 4(7):e1000118.

[40] Libby P, Ridker PM, Maseri A. Inflammation and atherosclerosis. Circulation. 2002; 105(9):1135-1143.

[41] Cídl K, Strelcová L, Znojil V, Váchi J. Angiotensin I-converting enzyme (ACE) polymorphism and ABO blood groups as factors codetermining plasma ACE activity. Exp Hematol 1996; 24(7):790-794.

[42] Garrison RJ, Havlik RJ, Harris RB, Feinleib M, Kannel WB, Padgett SJ. ABO blood group and cardiovacular disease: the Framingham study. Atherosclerosis 1976; 25(2-3):311-318.

[43] Garratty G. Relationship of blood groups to disease: do blood group antigens have a biological role? Rev Med Inst Mex Seguro Soc 2005; 43 (Suppl 1):113-121.

[44] Garratty G. Blood groups and disease: a historical perspective. Transfus Med Rev 2000; 14(4)291-301.

[45] Combs MR. Lewis blood group system review. Immunohematology 2009; 25(3):112-118.

[46] Hein HO, Sørensen H, Suadicani P, Gyntelberg F. The Lewis blood group--a new genetic marker of ischaemic heart disease. J Intern Med 1992; 232(6):481-487.

[47] Enevold C, Nielsen CH, Molbo D, Lund R, Bendtzen K, Fiehn N-, Holmstrup P. Lewis and AB0 blood group-phenotypes in periodontitis, cardiovascular disease, obesity and stroke. Sci Rep 2019; 9(1):6283.

[48] Natarajan R, Dhawan HK, Choudhury S, Vijayvergiya R, Marwaha N. Lewis blood group phenotype vis-a-vis biochemical and physiological parameters of coronary artery disease: A study in North Indian population. Asian J Transfus Sci 2020;14(1):9-12.

[49] Suadicani P, Hein HO, Gyntelberg F. Weight changes and risk of ischaemic heart disease for middle aged and elderly men. An 8-year follow-up in the Copenhagen Male study. J Cardiovasc Risk 1997; 4(1):25-32.

[50] Hein HO, Sørensen H, Suadicani P, Gyntelberg F. Alcohol consumption, Lewis phenotypes, and risk of ischaemic heart disease. Lancet 1993; 341(8842):392-396.

[51] Hein HO, Suadicani P, Gyntelberg F. Lewis phenotypes, leisure time physical activity, and risk of ischaemic heart disease: an 11 year follow up in the Copenhagen male study. Heart 2001; 85(2):159-164.

[52] Petit JM, Morvan Y, Mansuy-Collignon S, Viviani V, Vaillant G, Matejka G, Aho S, Guignier F, Brun JM. Hypertriglyceridaemia and Lewis (A-B-) phenotype in non-insulin-dependent diabetic patients. Diabetes Metab 1997;23(3):202-204.

[53] Höher G, Fiegenbaum M, Almeida S. Molecular basis of the Duffy blood group system. Blood Transfus 2018; 16(1):93-100.

[54] Kim S, Eliot M, Koestler DC, Wu WC, Kelsey KT. Association of Neutrophil-to-Lymphocyte Ratio With Mortality and Cardiovascular Disease in the Jackson Heart Study and Modification by the Duffy Antigen Variant. JAMA Cardiol 2018; 3(6):455-462.

[55] Afenyi-Annan A, Kail M, Combs MR, Orringer EP, Ashley-expression is associated with organ damage in patients with sickle cell disease. Transfusion 2008; 48(5):917-924.

Chapter 2

Human Blood

Francisca Varpit and Vela Galama

Abstract

The human blood is composed of 3 layers of humors when separated into its different components. The component that is clear, slightly yellow (plasma), the whitish viscous-like (buffy coat) and the red fluid (red cells). The plasma component is composed of proteins; however, it will not be discussed in this chapter. The buffy coat is composed of white blood cells and platelets. The white blood cells are composed of granulocytes and agranulocytes; all of which take part in immune defense. The granulocytes, including monocytes have non-specific immune response while agranulocytes, which include B and T cells have specific immune response. The platelets function to help maintain normal hemostasis during vascular injury. Blood group antigens are found on the surface of red cells and are composed of proteins, carbohydrates and lipids. They are mostly inherited on autosomes with the exception of two which have been found to be inherited on the X chromosomes. With the advance of technology, some of their physiological functional roles have been elucidated. These include; structural integrity, cationic exchange, transporters, adhesion and receptor functions, and cell to cell communication. However, these mechanisms have been capitalized by infectious agents to gain entry to the human body causing disease.

Keywords: human blood, blood components, red cell membranes, blood group systems, blood group antigens, physiological functions, pathophysiological functions

1. Introduction

Since ancient times, blood has been viewed as the very essence of human life. In fact, description of the human blood dates back to the writings of Hippocrates in about 400 B.C. He described it as being composed of four layers of fluid; one that looked "black bile, red blood, whitish viscous-like (phlegm) and yellow bile" [1]. This was later clarified in the twelfth century by a Swedish physician as a description of blood that is undergoing clotting process, where blood is being separated into distinct portions. During this time and earlier, the state of health and disease were thought to have been caused by an imbalance between these layers of blood. This helped to explain why bloodletting was performed during those ancient times and into the nineteenth century [1, 2].

During those early years (200 AD), red blood was taught to be the dominant humor and therefore bloodletting was carried out to reduce excess blood from circulation, to slow down the heart rate, and also to reduce an inflammatory process in an individual suffering from an inflammation. It was believed that this process would in turn bring balance to the different layers of blood, and ultimately health to the individual being bled [2].

The composition of blood was however not known, until the discovery of the microscope in 1673 by Anton Leeuwenhoeck. By 1683, Leeuwenhoeck could see minute microorganisms such as bacteria using the instrument [3] and even accurately described and measured red blood cells [1]. During this time, bloodletting was based on unscientific principles and therefore remained controversial.

However, with the advance of technology at the turn of the twentieth century, new approaches and standardized methods were developed, which contributed to better understanding of the composition and structural organization of red cells [4]. This eventuated in the current acceptance of Bloodletting as a therapeutic treatment for specific chronic diseases associated with hematochromatosis (iron overload), erythrocytosis (elevated hematocrit), porphyria and polycythemia (excess number of red cells) [3].

As a result of advance in technology, there was also the understanding that red blood cell membranes are composed of protein and lipid residues, which define their structural composition, physiological and biological roles. Much of their protein content is made up of hemoglobin; essential for transportation of oxygen (O_2) to tissues and carbondioxide (CO_2) to the lungs. Apart from transportation, they have other physiological functions such as; maintaining structural integrity of the cell, modulation of cell-cell interactions or vascular endothelium-cell interactions, anchorage site of cystoskeleton, anion exchange, adhesion and receptor functions [4]. It is now well understood that, apart from these normal physiological functions, red cells also serve as biological mediums by which external invaders can enter blood circulation and tissues to cause disease to the human body [5]. Their pathophysiological roles are discussed in Section 4, Subsection 4.2.

2. Blood Components

2.1 Composition

Blood does not only contain fluid but also other substances such as proteins, carbohydrates, lipids and antigens [4]. The blood is comprised of 55% plasma and 45% formed or cellular elements [6]. The formed or cellular elements include buffy coat, which contains white blood cells (WBCs) and platelets, and red blood cells (**Figure 1**). Blood is essential in humans and other animals and performs multiple tasks. The erythrocytes or red blood cells contain hemoglobin and function in the

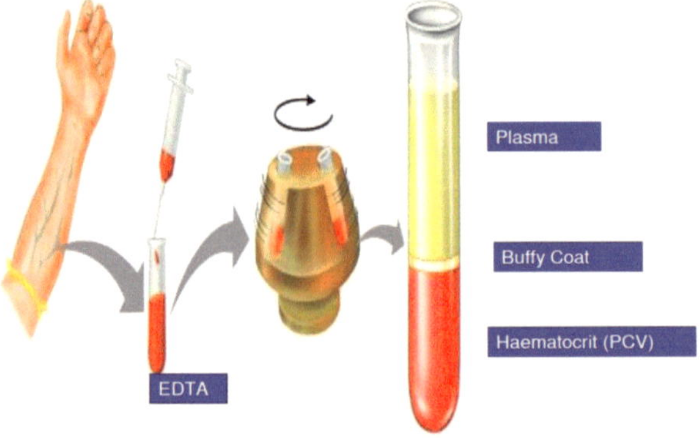

Figure 1.
Blood, when centrifuged, it is separated into 3 different layers; plasma (contains proteins), Buffy coat (contains mainly white blood cells and platelets), and red cell concentrate) [7].

transport of oxygen (O2) to tissues and carbon dioxide (CO_2) from tissues to the lungs. The WBCs (Leucocytes)) are involved in the body's defense against the invasion of foreign antigens. The Platelets (thrombocytes) are involved in the process of hemostasis; prevention of blood loss during injury [6–8].

2.2 Synthesis of blood components

All human blood goes through a process known as hemopoiesis or hematopoiesis. Hematopoiesis or hemopoiesis is the process of blood production. This process proceeds through different stages starting from early embryonic life (mesoblastic stage), to the hepatic stage, and then to the myeloid stage. The process in the embryonic and early foetal life occur in the yolk sac, although at this stage, only the very early red cells (erythroblasts) are formed. As the foetus develops, all blood cells are formed in the foetal liver. Apart from the yolk sac, hemopoiesis also occurs in the mesoderm of intra-embryonic aorta/gonad/mesonephros (AGM) region. After several weeks, it also occurs in the spleen, lymph nodes and thymus. From the third to the 4th month until birth, blood cell production occurs exclusively in the bone marrow. As the child matures into adulthood, blood cell production is confined to only the flat bones such as; the sternum, ribs, iliac bones, vertebrae, and proximal ends of long bones (**Figure 2**) [7, 8].

2.2.1 Red blood cells (Erythrocytes)

Mature red blood cells (RBCs) or erythrocytes are very small cells with a diameter of about 8.0 μm. They maintain a biconcave discoid shape and do not contain a nucleus [9]. The life span of mature RBCs is about 120 days. After the 120 days, they are phagocytosed by reticuloendothelial system (RES) macrophages in the spleen [8]. Their main function is to transport O_2 from the lungs to various tissues and organs and from these tissues and organs, it carries CO_2 to the lungs for reoxygenation [7, 9, 10].

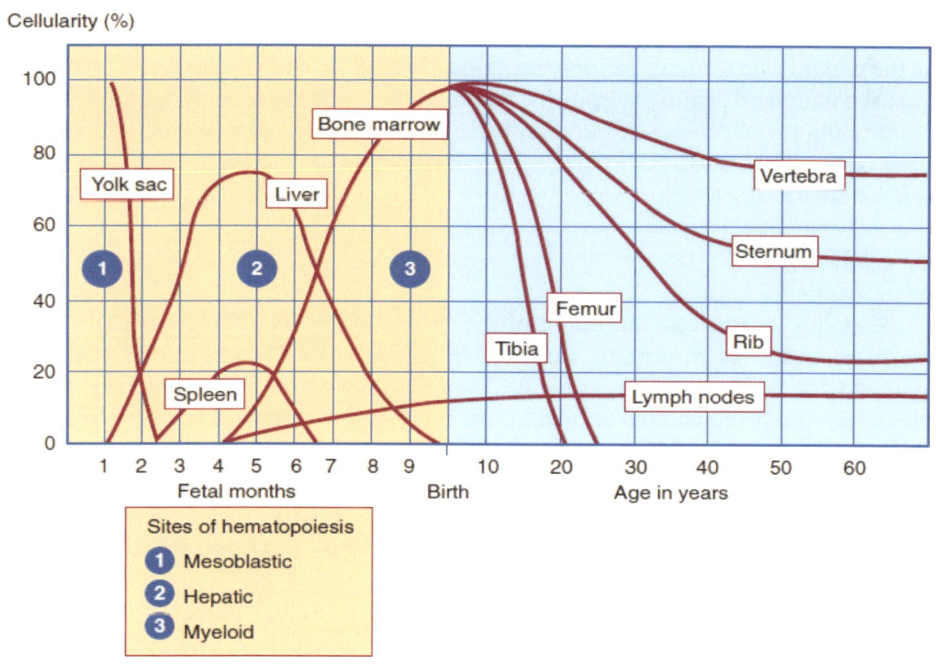

Figure 2.
The phases of hematopoeisis. During the process of blood production, erythroblasts are the first to be formed in the yolk sac, followed by the rest of the cell lines in the fetal liver, then in the spleen and finally in the BM from the 4th month of life onwards to adulthood [7].

2.2.2 White blood cells

White blood cells are categorised as granulocytes and agranulocytes. Granulocytes include; neutrophils, eosinophils and basophils, and agranulocytes include; lymphocytes and monocytes [9, 11].

2.2.2.1 Granulocytes

These are also called polymorphonuclear leucocytes because their nuclei are oddly shaped and their cytoplasm have densely stained granules when stained with Leishman's stain or Wright-Giemsa stain. They play a major role in immune defence as part of the innate immune system. Their role in defence is non-specific, short-lived and without memory [11].

2.2.2.1.1 Neutrophils

Neutrophils make up 97% of the granulocyte lineage and are the first to arrive at sites of infection. The size of a mature neutrophil is about 10 to 12 micrometres (μm), with 2 to 5 lobes of deep purple nucleoli when stained with Leishman's stain. They also have very fine light pink cytoplasmic granules. These granules contain proteins and enzymes such as; lysozyme, lactoferrin, vitamin B_{12}-binding protein, myeloperoxidase, acid phosphatase, elastases and others. Their major roles are; phagocytosis of bacteria, viruses and yeasts, formation of Neutrophil Extracellular Trap (NET), degranulation and cytokine production [11, 12]. Their lifespan is 1–2 days in peripheral blood circulation [7, 8].

2.2.2.1.2 Eosinophils

Eosinophils are small bilobed granulocytes, with granules that stain red orange with the Wright-Giemsa stain. These granules contain proteins and enzymes such as the major basic protein, cationic proteins, peroxidase and histaminase. These are used to defend against helminthic parasites. On activation, eosinophils produce debilitating toxic respiratory burst and also create transmembrane plug that kill their target. Their maturation in the BM takes 2 to 6 days and their lifespan in blood is less than 8 hours [7–9].

2.2.2.1.3 Basophils

Basophils are the least numerous of the circulating WBCs. Their nucleus contains condensed chromatin, shrouded by darkly stained coarse granules when stained with the Wright-Giemsa stain. These granules contain inflammatory mediators and proteins such as; histamine, serotonin, heparin, Major Basic protein, and enzymes such as DNAases, proteases and lipases. They also express receptors to IgE and therefore have the ability to become activated when bound to IgE-Antigen immune complexes. On activation, they degranulate releasing their content that kill their target [12]. They also play a role in hypersensitivity reaction. Their lifespan is less than 3 days [13].

2.2.2.2 Agranulocytes

These cells do not contain multiple lobes like the granulocytes.

2.2.2.2.1 Lymphocytes

There are two types of lymphocytes. These are B and T lymphocytes. They are the major players in the adaptive immune response against foreign invasions. They constitute 20–30% of the total WBC population. Unlike the granulocytes, their actions are slow, specific and have memory [12]. The B cells function to produce antibodies and the T cells function to provide help to B-cells for antibody production, kill off virus-infected cells, and also play regulatory roles. The sizes of these cells range from 8 to 10 μm [7–9]. Naïve lymphocytes live longer in their restful state than effector lymphocytes. Lymphocytes that have differentiated into memory cells have longer lifespan. The different lifespan periods are dependent on heterogeneous populations during stages of differentiation and activation [14].

2.2.2.2.2 Monocytes

Monocytes are usually large in size, measuring about 12–20 μm in diameter. The nucleus is generally kidney-shaped with fine chromatin. They have abundant cytoplasm which appear blue, and often contain azurophilic granules and vacuoles. They circulate in blood for about a day before they exit to tissues where they are called macrophages or histiocytes [12]. In the blood, their function is to protect against blood-borne pathogens. In tissues, their major roles are phagocytosis, antigen presentation, cytokine production and NET formation [7–9]. This population of cells comprise a heterogeneous population; distinguishable by their cell surface markers and functions. A blood classical monocyte's lifespan is ~1.0 day, a blood intermediate monocyte's life span is ~ 4.3 days and a non-classical blood monocyte's lifespan is ~7.4 days [15].

2.2.3 Platelets

Platelets are the products of cytoplasmic fragmentation of megakaryocytes in the BM in a process called megakaryopoiesis [12]. The diameter of platelets is about 2–3 μm. They contain α-granules, dense granules, and lysosome. Their main function is their synergistic interactions with endothelial wall and plasma proteins to maintain normal hemostasis during vascular injury. They remain alive in the blood for about 10 days [7–9].

3. Blood group antigens

The term "blood group" generally refers to a person's collection of "red cell surface antigens". These are found on the surface of red blood cells as part of the red cell membrane. They are made up of proteins, glycoproteins and glycolipids. These may elicit an immune response in individuals lacking these antigens.

Blood group antigens have been classified by the International Society for Blood Transfusion (ISBT) into 30 blood group systems [16–18]. However, as of June, 2019, there are currently 38 blood group systems (**Table 1**), with over 200 red cell antigens classified under these systems, while some are classified as collections [16, 19]. These antigens are inherited and as such, understanding of their genetic makeup is important as well as their unique characteristics which differentiates one group or one antigen from another. Such characteristics include; genetic expression, structure and location on red blood cells, and the type of antibody they induce. The most common and clinically significant of these blood group antigens are the

Traditional name	ISBT No.	Symbol	Traditional name	ISBT No.	Symbol
ABO	001	ABO	Gerbich	020	GE
MNS	002	MNS	Crower	021	CROM
P1Pk	003	P1	Knobs	022	KN
Rh	004	RHD, RHCE	Indian	023	IN
Lutheran	005	LU	Ok	024	OK
Kell	006	KEL	Raph	025	RAPH
Lewis	007	Le	John Milton Hagen	026	JMH
Duffy	008	FY	I	027	I
Kidd	009	JK	Globoside	028	GLOB
Diego	010	DI	GIL	029	GIL
Yt	011	YT	RHAG	030	RHAG
Xg	012	XG	Forssman	031	Fors
Scianna	013	SC	Jr	032	Jr^a
Dombrock	014	DO	Lan	033	Lan
Colton	015	CO	Vel	034	Vel
Lansteiner–Wiener	016	LW	CD59	035	CD59.1
Chido/Rodgers	017	CH/RG	Aug	036	Aug1
Hh	018	H	Kanno	037	Kanno
Kx	019	XK	Sid	038	Sid

Table 1.
International society for blood banks (ISBT) classification of the known blood group systems with their ISBT numbers and symbols [19].

antigens belonging to the ABO and Rhesus blood group systems because they have the potential of eliciting an immune response that can cause fatal consequences through blood transfusion and or pregnancy [16].

3.1 Blood group antigen inheritance

Genes are units of inheritance that encode particular proteins needed for production of particular inherited traits. Genetic information caried in these genes are found in double stranded deoxyribonucleic acid (DNA) called chromosomes. Humans have 23 pairs of these, of which; 22 are autosomes and 1 pair of sex chromosomes. Within these chromosomes are sites called genetic loci; where genes are located. Within these genetic loci are alternate forms of genes called "alleles" [16].

Inheritance of blood group antigens follow an autosomal codominant pattern of inheritance, where alleles inherited from both parents are equally expressed on autosomes [16]. Hereditary patterns of these antigens are based on the Mendelian principles of inheritance; which sprouted from early experiments done on pea hybrids [17]. His genetic concepts of Dominance, Independent Segregation and Assortment are currently applied in understanding inherited characteristics (traits) observed in human blood group genetics. For each inherited trait (character), there are two alleles inherited from each parent. The expression of this inherited trait is dependent on the combination of the two alleles inherited. One of these alleles suppresses the effect of the other, while the other allele can only be observed in the absence of the dominant allele. The allele that suppresses the expression of the other

Parents	A	A
A	AA	Aa
a	aA	Aa

Table 2.
This Punnett square describes Mendel's concept of "independent segregation" using symbolic "A" and "a" to denote inheritance of a dominant allele and a recessive allele respectively [17].

is called the dominant gene, while the other that is not expressed in the presence of the dominant gene is called recessive. This concept is referred to as the Law of dominance [16, 17].

Thus, alleles inherited from each parent can be the same (homozygous) or different (heterozygous). The concept of independent segregation refers to each parent having a set of alleles for a particular trait, either of which can be passed onto the next generation. These alleles segregate, allowing for only one allele to be transmitted to an offspring. For example, using the letters "A" and "a" to represent a dominant and a recessive allele respectively, there are four possible combinations (**Table 2**). The AA combination constitutes entirely of AA (homozygous) offspring and the aa combination comprise entirely of aa (recessive) offspring. The offspring of the aa allele combination differ from the AA, aA and Aa combination due to the absence of a dominant allele. The offspring of Aa and aA (heterozygous) gene combinations inherited traits common to all four combinations [17].

Mendel's third concept is "independent assortment". During meiosis, a mixture of genetic material is produced resulting from random behavior of genes on separate chromosomes. These genes are inherited independent of each other on different chromosomes but are expressed on the same red cell membrane. **Figure 3** illustrates

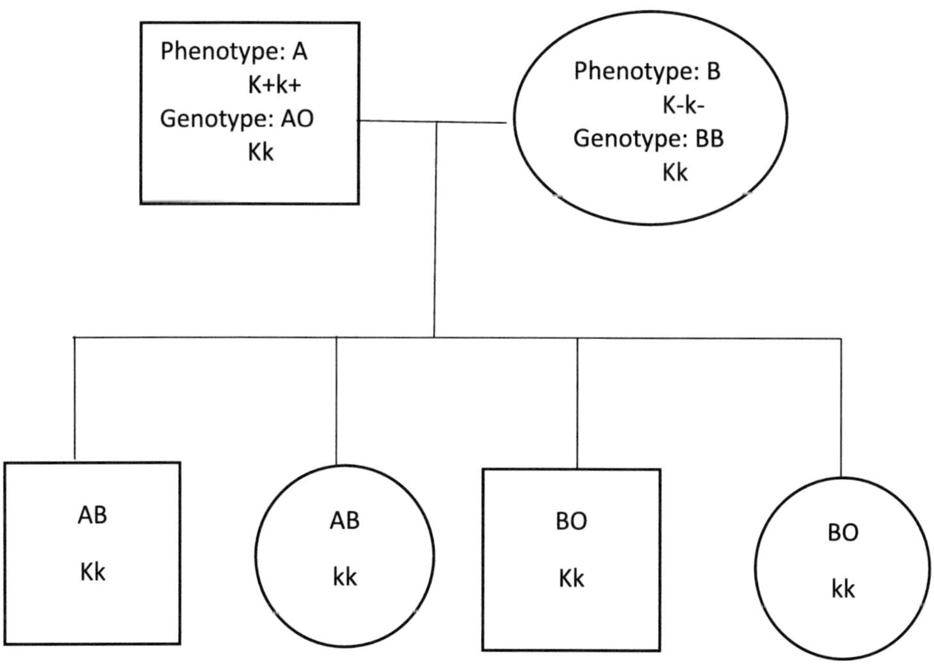

Figure 3.
Independent assortment. The ABO blood group antigens are sorted independently from the Kell antigen genes because they are inherited on different chromosomes. However, they can all be expressed on the same red cell membrane separately and discretely [16].

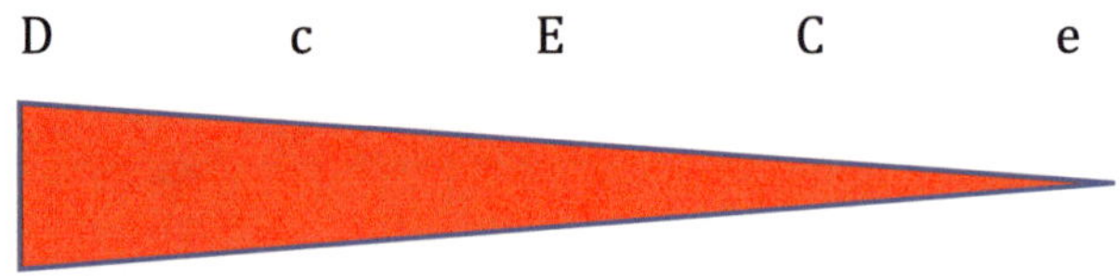

Figure 4.
Strength of immunogenicity of the Rhesus Blood Group antigens in decreasing order of immunogenicity.

the ABO and the Kell blood group system genes, whose genes are located on chromosomes 9 and chromosome 7 respectively [16].

Genes coding for most of these blood group antigens are inherited on autosomes except the Xg and the Kx. The Kx is coded for by the Xk gene, while the Xg is coded for by the Xg^a allele, both of which are located on the X chromosome. This means that fathers having the latter genes would pass it on to their daughters and none to their sons. If, however, mothers have these genes, they would pass it on to both genders [16, 20].

Inheritance of blood group antigens follow distinct patterns of inheritance. Some genes code for antigens that are co-dominant., some are dominant over another and some are recessive. For example, in the ABO blood group system, the A and B antigens are codominant; both alleles are expressed to show the trait. When A and O alleles, or B and O alleles are inherited, the O trait is not expressed because the A and the B alleles are dominant over the O. When O and O alleles are inherited, the O traits are observed in the absence of a dominant gene. The O gene is said to be recessive (**Figure 4**) [17].

According to the ISBT, some of these blood group antigens have been classified as blood group systems based on their serologic and molecular characteristics (**Table 1**) and their locations on specific chromosomes has also been elucidated (**Table 3**).

3.2 Structure and location on red cells

Three of the basic structural properties of red cells are hemoglobin, enzymes and the membrane [18].

3.2.1 Hemoglobin

Although mature red cells no longer have nucleus nor mitochondria, they have an abundance of hemoglobin, a red pigment that contains oxygen. The hemoglobin carries oxygen to all parts of the body to keep the body alive and collects carbon dioxide (CO_2) from the tissues to the lungs for re-oxygenation.

Blood group system	Chromosome	Blood group system	Chromosome
Rh	1	Kidd	18
Duffy	1	Lewis	19
MNS	4	Landsteiner–Wiener	19
Chido/Rodgers	6	Lutheran	19
Kell	7	Hh	19
ABO	9	P	22
Kx	X	Xg	X

Table 3.
Some of the blood group antigens and their chromosomal locations, all of which are found on autosomes and only two are found on the X (sex) chromosomes [16].

3.2.2 Enzymes

The Embden-Meyerhof pathway is a metabolic pathway used by mature red cells to generate energy through a series of enzymatic pathways that catalyze the conversion of glucose to lactate and pyruvate. Within this pathway, there is a shunt called the Rapoport Luebering shunt that generates the production of 2,3Diphosphoglycerate (2,3DPG); important in influencing the release of O_2 in tissues. Apart from production of energy and 2,3 DPG, another end product of this glycolytic pathway is generation of Nicotinamde Adenine Dihydrogenase (NADH); necessary for reducing nonfunctional methemoglobin to oxyhemoglobin. Another metabolic pathway in red cells is the pentose phosphate shunt. Two enzymes are generated during this process. These are called glucose-6 phosphate-dehydrogenase (G6PD) and 6-phosphogluconate dehydrogenase (6PGD), which then generate Nicotinamide Adenine Diphosphodehydrogenase (NADPH). This enzyme with Glutathione reductase catalyzes the generation of glutathione (GSH), which along with glutathione peroxidase function to detoxify hydrogen peroxide and thus rendering the red cell safe from oxidant damage [18].

3.2.3 Red cell membrane

The red cell membrane is comprised of proteins, lipids and carbohydrates. Blood group antigens are found linked to these on the red cell membranes. Some of these red cell antigens are only found on the red cell membrane or in fluids, while some are found both on the red cell membranes and in body fluids. Those found on the red cell membrane and also in fluids are; the ABO blood group system, Hh and Lutheran. In fact, the ABO antigens are also found in lymphocytes, platelets, epithelial cells and endothelial cells and the kidney having been adsorbed from the plasma [16]. Those found on red cell membranes include; the Rhesus, Kell, Kx, Duffy and Kidd. There is one known blood group system that is unique from the others in that it is found predominantly in fluids. This blood group system is called the Lewis and it is produced by tissue cells and released into body fluids. However, these get adsorbed onto red cell membranes shortly after birth and continues on for the first 6 years of life [19]. Although it is not clinically significant, serologic antibody detected may indicate H. *pylori* infection as the Le^b allele has a receptor to this gram- negative bacteria [16].

Like the ABO blood group system, the Le gene does not code directly for its antigen but instead codes for a glycosyltransferase called L-fucosyltransferase, which then adds an immunodominant sugar called L-fucose to the precursor substance (H antigen) to form the Le^a antigen or Le (a+) phenotype. Its adsorption onto red cell membranes depends on the presence of the precursor substance (H antigens) on red cells and also presence of the L-fucosyltransferase type 1. Conversion of the Lewis antigen from Le^a allele to Le^b in secretions depends on the Secretor gene (Se). If the individual also had inherited this gene, which is mainly found in fluids, then the L-fucose is added to the precursor substance in fluids to form the Le^b antigen or Le(b+) phenotype. This then gets adsorbed onto red cell membrane preferably over the Le^a antigen [16, 20].

3.2.3.1 The ABO, rhesus and other blood group antigens

Genes coding for the formation of some of these antigens do not code directly for their respective antigens. Instead, they code for glycosyltransferases, which in turn catalyze the transfer of immunodominant sugars from donor molecules to a precursor substance for the formation of their respective antigens. Examples of these are; the H, ABO, Se, Lewis, I/i and P_1. The H, ABO and secretor (Se) blood group antigens are inter-related, in that the H antigens forms the basis for the formation of the A and B antigens on red cells and in secretions [20]. Lack of formation of the H precursor antigen on red cells result in the formation of the Bombay (OH) blood type, instead of the ABO [21]. These individuals lack the ABH antigens and thus possess naturally occurring antibodies against the A, B, O and H antigens in their plasma. This exposes them to fatal consequences during blood transfusion. The Se gene on the other hand play a vital role in the formation of ABH antigens in secretions. This is because it controls the expression of the H antigen in secretions [21]. In individuals who lack the Se genes, there is no formation of the H antigen in secretion and ultimately no formation of the A and B antigens in secretions as well.

There are however, cases where an individual inherits a dominant allele and a recessive (Sese) or co-dominant alleles (SeSe) from each parent in secretions but lacks the H gene on red cells. In such individuals, formation of the H antigen will still occur and thence formation of A and B antigens in secretions but not on red blood cells. These individuals are said to have the "para-Bombay" phenotype [22]. Some amounts of the A and B antigens may adsorb onto red cell membranes from plasma and hence are detected on red blood cells [22]. This is summarized in **Table 4**.

Some of the blood group antigen genes that code directly for the formation of their respective antigens include the Rhesus, Kell, Kidd, Duffy and MNS. Among these, the Rhesus antigens are very immunogenic because they are protein in nature, the most immunogenic of which is the Rh D, followed by the c, E, C, e. (**Figure 4**). Initially founded in 1939, the Rh blood group system is the most complex and polymorphic with about 50 well-defined related antigens assigned to its system classification by the ISBT [8, 23]. It is mostly associated with Haemolytic Disease of the Foetus/New Born (HDFN) especially in a second pregnancy of a Rh D positive child conceived in a mother who does not have the Rh D antigen. In the general population, 85% have the Rhesus D antigen, while the rest are negative for it [19, 24].

However, the Kidd, Kell and Duffy are all considered clinically significant as well because they have also been implicated in Hemolytic Disease of the Newborn and therefore recognition of antibodies against these antigens are vital in relation to blood transfusion and pregnancy [16].

After the Rhesus antigens, the Kell blood group antigens are next most immunogenic. It is associated with another blood group system called the Kx, inherited on the X chromosome. Absence of this on an individual's red cells weakens the

Inherited	Genes		Antigenic expression	
			On red cells	In saliva (secretions)
AB	HH	SeSe	A, B, H	A, B, H
AB	HH	sese	A, B, H	None
OO	HH	SeSe	H	H
OO	HH	sese	H	None
oh	h	sese	None	None
para-oh	h	SeSe	None	H, some amounts of A, B antigens adsorbed from plasma
para-oh	h	Sese	None	H, some amounts of A, B antigens adsorbed from plasma

Table 4.
Interaction between the ABO, H and Secretor genes depicting the expression of soluble antigens on red blood cells and in secretions [16, 22].

expression of the Kell antigens, a condition called "Mcleod phenotype". They develop red cell abnormalities such as acanthocytosis and reticulocytosis [16, 20].

The Duffy Blood group antigens are glycoproteins found on chromosome 1. These glycoproteins span the membrane of the red cells. First defined in 1950 in a patient who was suffering from hemophilia, it is best remembered for its association with malaria. Individuals who do not possess this antigen (Fy a-b-) are protected against Plasmodium vivax and Plasmodium knowlesi infections. This blood type is common amongst African and American Blacks [16, 25]. Like the Duffy, the Kidd blood group antigens are glycoproteins but located on chromosome 18. They are not as polymorphic as the Rhesus and Kell blood group antigens. They play a role in urea transport. They have also been implicated in causing extravascular haemolysis in a delayed type of hemolytic transfusion reaction [16, 20].

Like the ABO and Lewis blood group antigens, the I blood group antigens are oligosaccharides, which along with the i antigen, exist on the ABH oligosaccharide chain precursors nearer to the red cell membrane. The i antigen on the other hand has not been assigned to a blood group system and remains as a collection. The i antigen is mostly expressed on red cells of New Born and cord blood, while the I is mostly seen in adults red cells [16, 18]. The P1PK Blood group antigens are glycoproteins and glycolipids, and like the ABO, Lewis, and I blood group antigens, they are also formed through the actions of glycosyltransferases. At birth, P1 is poorly expressed. The MNS are structurally glycoproteins; their sugar components are primarily composed of sialic acids attached to proteins which lends the negative charge of red cells [20].

3.3 Antibody response to blood group antigens

Karl Landsteiner's discovery of the ABO blood group antigens in 1900 [16, 26] was the beginning of safe blood transfusion as we know today. He began by first experimenting with his own blood and then that of his co-workers. When he began to mix serum taken from co-worker A with red cells from co-worker B, he realized that these formed clumps. When he then mixed his serum with red cells from both of his two co-workers, he recognized that he had antibodies against both. This he appropriately called blood group O, which to become the universal donor. The other two co-workers A and B, he called anti-A and anti-B respectively because their serum agglutinated when mixed with each other's red cells. In his article published in 1900 on these experiments, he added a footnote that stated "the serum of healthy humans has an agglutinating effect, not only upon animal blood cells, but frequently upon blood cells from other individuals as well" [26].

The antibodies against the ABO blood group antigens are the most clinically significant because they are pre-existing. Based on Landsteiner's rule, healthy individuals have antibodies against antigens that they do not have. This is the basis for all blood transfusions today. Patients' blood is always typed and crossmatched before they are infused to avoid fatal intravascular hemolytic transfusion reactions. An individual with blood antigen A has antibodies against the B antigen and an individual with blood antigen B on their red cell membrane has pre-existing antibodies A in their plasma. Individuals with no AB antigens on their red cells have both A and B antibodies in their plasma, while those that have both A and B antigens on their red cells, they do not have pre-existing antibodies in their plasma (**Table 5**).

Antibodies against the ABO blood group antigens are mainly of the IgM class and therefore are capable of reacting at temperatures ranging from 4°C to 22°C or room temperatures. Because of its pentameric structure, it is able to bind to 10 of these red cell antigens in at any one time. This is enough to trigger off a massive intravascular complement protein reaction via the classical pathway resulting in an

Blood type (antigens)	Antibody-A	Antibody-B	Antibodies-AB
On red cells	**In plasma**	**In plasma**	**In plasma**
A	NIL	Present	Present
B	Present	NIL	Present
AB	NIL	NIL	NIL
O	Present	Present	Present

Table 5.
Based on Karl Landsteiner's conclusions, healthy individuals have antibodies in their plasma against antigens that they do not have.

acute hemolytic transfusion reaction if the wrong ABO blood is transfused to an individual with a different blood type. This type of reaction usually occurs within minutes or hours of transfusion of the wrong blood. If not stopped quickly, fatal consequences like disseminated intravascular coagulation (DIC), irreversible shock and death can occur. Antibodies against the Lu^b antigens in the Lutheran Blood group system are also clinically significant. Although rare and mainly of the IgG class, it has been found to be associated with transfusion reactions and HDFN [16].

Auto antibodies against the P antigen is bi-phasic and it is called the Donath-Landsteiner antibody. It is referred to as bi-phasic because it is able to bind to the P1 or P2 antigens at lower temperatures especially in the extremities of the body and when warmed to warmer temperatures, the complement cascade is activated resulting in haemolysis. These antibodies are mainly associated with paroxysmal cold hemoglobinuria; a rare disorder characterized by cold associated haemolysis and hematuria [16].

Autoantibodies against the I are mainly associated with patients with Mycoplasma pneumoniae infections and cold haemagglutinin disease. In these patients, strong auto-agglutinations are observed in in-vitro analysis. In patients with disease conditions such infectious mononucleosis, lymphoproliferative disease and sometime in cold hemagglutinin disease, anti-i is usually detected.

4. Blood group antigens as modes of disease transmission

For hundreds of years, the red cells were thought to be inert; with no form of biological functions. This was never been justified until in 1865, when Hoppe-Seyler discovered that in the red cells, there is an abundance of hemoglobin. These red cells have two important properties that allow them to squeeze easily through blood capillaries supplying tissues with oxygen to keep these tissues alive [1]. These are flexibility of its membrane and fluidity of its content. Any imbalance in these two properties will cause reduced survival of these red cells and removal hence by macrophages in the spleen [20].

4.1 Physiological functions of red cell antigens

Apart from the important roles in O_2 transport to tissues and CO_2 back to the lungs, production of ATP, 2,3 DPG and production of enzymes that ultimately catalyzes the biochemical processes that result in the reduction of the dysfunctional methemoglobin to oxyhemoglobin and reduced GSH, red cell antigens located on and across the red cell membranes also play other roles in many ways. Some of their known physiological functions include, gylcosyltransferases, structural maintenance of red cells, protein transportation, complement pathway molecules

Blood group antigens	Putative functions on red cell membrane	Blood group antigens	Putative functions on red cell membrane
ABO	Glycosyltransferases	Chido/Rodgers	C4 complements adsorbed onto red cell membrane
MNS	Contributes to glycocalyx. GPA likely acts as a chaperone for Band 3	Hh	Fucosyltransferase
P	Glycosyltransferase	Kx	Xk protein linked to Kell glycoprotein. Homology to neurotransmitter transporters.
Rh	Involved in CO_2/O_2 or NH_4^+/NH_3 transport or maintenance of cell shape	Gerbich	Glycophorins C & D. Could link to glycocalyx. Links membrane to cytoskeleton
Lutheran	Binds laminin 511 and 521. Probably adhesion/receptor in erythropoiesis	Cromer	Decay-accelerating factor. Inhibits activity of C3 convertase. Also protects cell from lysis by autologous complement
Kell	Possibly processes endothelin 3	Knops	Complement receptor 1. Binds and processes immune complexes
Lewis	Not synthesized on red cell membranes	Indian	Binds hyaluronan. Probably adhesion/receptor
Duffy	Antigen receptor for chemokines, possibly used for removal from peripheral blood	Ok	Probably adhesion/receptor
Kidd	Urea transporter	Raph	May associate with integrins to generate laminin-binding complexes
Diego	Band 3 anion exchanger 1. Exchanges HCO_3^-/Cl-. Links membrane to cytoskeleton	John Milton Hagen	Probably adhesion/receptor
Yt	Its function is unknown	I	Glycosyltransferase involved in branching of oligosaccharide chains
Xg	Possibly adhesion/receptor	Globoside	Galactosytransferase
Scianna	Possibly adhesion/receptor	GIL	Water and glycerol channel
Landsteiner–Wiener	Intracellular adhesion molecule-4. Binds integrin. Possibly adhesion /receptor involved in stability of erythroblastic islands	RHAG	Rh associated glycoprotein probably involved in CO_2/O_2 or NH_4^+/NH_3 transport

Table 6.
Putative physiological functions of some of the known red cell antigens [22].

regulation, adhesion molecules and as microbial receptors [16, 23]. These functions are summarized in **Table 6**.

4.2 Pathophysiological functions of blood group antigens

Blood is a pharmaceutical therapy for treatment of various blood component deficiencies and blood loss. However, it is quite often forgotten that, apart from its normal biological functions, it also serves as vessel for transmission of various blood borne pathogens. It is now well documented that some blood group antigens have been found to be associated with increased susceptibility to infections [27] enhance disease progression in others [28], while in others have indicated reduced susceptibility and severity [29]. Susceptibility to infection often depends on the geography and epidemiology of the different blood group antigens [30].

Among the ABO blood group antigens, the A antigen has been shown to be associated with increased mortality from the COVID-19 than Blood groups B and O [30]. With Blood groups B and AB, they have a higher risk of suffering from thromboembolism caused by the COVID-19 infections than O because they have higher levels of von Willie Brands Factor (vWF) [31]. Blood group A has also been reported to play a synergistic effect with the Hepatitis B virus (HBV) on the risk of development of pancreatic cancer [32]. Blood group A is also found to be significantly associated with the HBV infections, while syphilis was significantly associated with the Rhesus blood group in the same study [33]. Data from one study demonstrated that Blood group B antigens are associated with lower risk of being infected with hepatitis B virus (HBV) while Group O has been demonstrated to have had a 12% risk of being infected with HBV [32]. Severity of diarrhoea caused by *Esherishia coli, Vibrio Cholerae and Helicobacter pylori* is dependent on the O blood type and Secretor status of an individual [34]. Blood group AB is associated with the severity of dengue disease in secondary infections [27]. Like the covid-19 infections, blood group O plays a protective role against severe malaria infection [35, 36]. Being a non-secretor also play a role in reducing risk of infection by the HIV-1 and also slows down disease progression. However, on the other extreme, being a Secretor promotes infections by *Haemophilus inflenzae, Neisseria meningitidis, Streptococcus pneumonie and Esherishia coli* [34].

Helicobacter *pylori* is associated with peptic ulcer, disease, gastric carcinoma and the Norwalk virus, however disease progression is enhanced in the presence of Le^b antigens as these serve as receptors to the bacteria [16]. Furthermore, autoantibodies against the I (autoanti-I) are increased in *Mycoplasma pneumoniae* infections and Cold haemagglutinin disease. Additionally, autoantibodies against the i antigens (autoanti-i) are elevated in infectious mononucleosis, lymphoproliferative disease and also in Cold Hemagglutinin disease. The Duffy blood group antigens on the other hand play a protective role against Plasmodium vivax (Pv) invasion [24]; a parasite species that causes malaria. This applies only to individuals who does not express the Duffy antigens (Fya-b-).

5. Conclusion

The human blood is the very essence of life as it supplies the whole body with O_2 and nutrients needed for its sustenance of life. The blood group antigens are a part of this scared suspension of fluid that flows throughout the body unendingly throughout life. Carried in its membranes are the structures that can serve two purposes in a normal physiological sense and pathophysiological, in that apart from carrying out functions that sustains the livelihood of the body that carries it, it also serves as a means of entry for foreign invaders, which cause an imbalance in the normal physiological functioning of the body causing disease state.

Acknowledgements

We wish to acknowledge the contribution of Mr. Gairo Gerega of the University of Papua New Guinea School of Medicine and Health Sciences for supplying some notes on the blood group antigens.

Conflict of interest

"The authors declare no conflict of interest."

Author details

Francisca Varpit* and Vela Galama
Division of Health Sciences, Discipline of Medical Laboratory Science, University of Papua New Guinea School of Medicine and Health Sciences, Port Moresby, Papua New Guinea

*Address all correspondence to: lakuadavina@gmail.com

References

[1] McKenzie SB. Text Book of Haematology. Philadelphia, PA, USA: Lea & Febiger; 1988. p. 507

[2] DePalma RG, Hayes VW, Zacharski LR. Bloodletting: Past and present. American College of Surgeons. 2007;**205**(1):132-144. DOI: 10.1016/j.jamcollsurg.2007.01.071

[3] Singer C. Section of the History of Medicine. Notes on the History of Microscopy. Thousand Oaks, CA: SAGE Publications; 2016. pp. 247-279

[4] de Oliveira S, Saldhana C. An overview about erythrocyte membrane. Clinical Hematology and Microcirulation. 2010;**44**:63-74. DOI: 10.3233/CH-2010-1253

[5] Marsh LW. Biological roles of blood group antigens. The Yale Journal of Biology and Medicine. 1990;**63**:455-460

[6] Brunning RD. Hematology: Basic principles and practice. JAMA: The Journal of the American Medical Association. 1995;**274**(21):1722-1723

[7] Hoffman R, Silbertstein LE, Anatasi J, Benz EJ Jr, Heslop HE, Weitz JI, et al. Hematology: Basic principles and practice. Amsterdam, Netherlands: Elsevier; 2017

[8] Hoffbrand AV, Pettit JE, Moss PAH. Essential Haematology. Oxford: Blackwell Science; 2001

[9] Marieb EN. Human anatomy and physiology. In: Human Anatomy and Physiology. London: Pearson; 2004. pp. 644-674

[10] Addai-Mensah O et al. Determination of haematological reference ranges in healthy adults in three regions in Ghana. BioMed Research International. 2019;**2019**: 7467512-7467512

[11] Abbas AK, Lichtman AH, Pillai S, editors. Cellular & Molecular Immunology. 8th ed. Philadelphia: Elsevier Saunders; 2015. p. 835

[12] Delves PJ, Martin SJ, Burton DR, Roitt IM, editors. Roitt's Essential Immunology. 13th ed. London: Willey-Blackwell; 2017. p. 556

[13] Yoshinori Y, Hajime K. Basophilis and mast cells in immunity and inflammation. Seminars in Immunopathology. 2016;**38**:535-537. DOI: 10.1007/s00281-016-0582-0

[14] Tough DF, Sprent J. Lifespan of lymphocytes. Immunologic Research. 1995;**14**:1-12

[15] Patel AA, Ginhoux F, Yona S. Review. Monocytes, macrophages, dendritic cells and neutrophils: An update on lifespan kinetics in health and disease. Immunology. 2021;**163**:261. DOI: 10.1111/imm.13320

[16] Blaney KD, Howard PR, editors. Basic & Applied Concepts of Blood Banking and Transfusion Practices. Amsterdam, Netherlands: Elsevier; 2013. p. 387

[17] Schwarzbach E, Smýkal P, Dostál O, Jarkovská M, Valova S. Gregor J. Mendel-genetics founding father. Czech Journal of Genetics and Plant Breeding. 2014;**50**(2):43-51

[18] Kawthalkar SM, editor. Essentials of Haematology. 2nd ed. India: Jaypee Brothers Medical Publishers (P) Ltd; 2013. p. 546

[19] Smart E, Armstrong B. Blood group systems. ISBT Science Series. 2020;**15**:123-150. DOI: 10.1111/voxs.12593

[20] Reid ME, Lomas-Francis C, editors. The Blood Group Antigen FactsBook. 2nd ed. New York: Academic Press; 2004. p. 552

[21] Kelly RJ, Ernst LK, Larsent RD, Bryant JG, Robinson JS, Lowe JB. Molecular basis for H blood group deficiency in Bombay (Oh) and para-Bombay individuals. Proceedings of the National Academy of Sciences of the United States of America. 1994;**91**: 5843-5847

[22] Subramaniyan R. Letter to the Editor. AB para-Bombay phenotype: A rare blood group variant and its clinical significance. Hematology, Transfusion and Cell Therapy. 2018;**40**(1):96-97

[23] Daniels G. The molecular biology of blood groups. Review. ISBT Science Series. 2009;**4**:368-374

[24] Cartron J-P, Cherif-Zahar B, Le Van Kim C, Mouro-Chankeloup I. The two-gene model of the RH blood-group locus. Biochemical Journal. 1995;**306**:877-878

[25] Nichols ME, Rubinstein P, Barnwell J, De Cordoba SR, Rosenfield RE. A new human duffy blood group specificity defined by a murine monoclonal antibody immunogenetics and association with susceptibility to plasmodium vivax. Journal of Experimental Medicine. 1987;**166**:776-785

[26] Tan SY, Graham C. Karl landsteiner (1868-1943): Originator of ABO blood classification. Singapore Medical Journal;**201**:243-244. DOI: 10.11622/smedj.2013099

[27] Abdollahi A, Mahmoudi-Aliabadi M, Mehrtash V, Jafarzadeh B, Salehi M. The novel coronavirus SARS-Cov-2 Vulnerability Association with the ABO/Rh Blood Types. Iran Journal Pathology. 2020;**15**(3):156-160. DOI: 10.30699/ijp.2020.125135.2367

[28] Kalayanarooj S, Gibbons RV, Vaunghn D, Green S, Nisalak A, Jarman RG, et al. Blood group AB is associated with increased risk for severe dengue disease in secondary infections. Journal of Infectious diseases. 2007;**195**: 1014-1017. DOI: 10.1066/512244

[29] Alkout TA, Alkout AM. ABO blood groups among coronavirus disease in 2019 patients. Iberoamarican Journal of Medicine. 2020;**2**:1-7. DOI: 10.5281/zenodo.3893256

[30] Anstee DJ. The relationship between blood groups and disease. The American society of Haematology. Blood. 2010;**115**(23):4635-4643. DOI: 10.1182/blood-2010-01-261859

[31] Le Pendu J, Breiman A, Rocher J, Dion M, Ruvoën-Clouet N. ABO blood types and COVID-19: Spurious, anecdotal, or truly important relationships? A Reasoned Review of Available Data. Viruses. 2021;**13**(160): 1-20. DOI: 10.3390/v13020160

[32] Jing W, Zhao S, Liu J, Liu M. ABO blood groups and hepatitis B virus infection: A systematic review and meta-analysis. BMJ. 2020;**10**:1-9. DOI: 10.1136/BMJOPEN-2019-034114

[33] Wang D-S, Chen D-L, Ren C, Wang Z-Q, Qiu M-Z, Luo H-Y, et al. ABO blood group, hepatitis B viral infection and risk of pancreatic cancer. International Journal of Cancer. 2012;**131**:461-468

[34] Batool Z, Durrani SH, Tariq S. Association of ABO and Rh blood group types to hepatitis B, hepatitis C, HIV and syphilis infection, a five year' experience in healthy blood donors in a Tertiary care Hospital. Journal of Ayub Medical Colleage Abbottabad. 2017;**29**(1):90-92

[35] Blackwell C, Dundas S, James VS, Mackenzie DAC, Brown JM, Alkaout AM, et al. Blood group and susceptibility to disease caused by Escherishia coli 057. Journal of Infectious Disease. 2002;**185**:393-396

[36] Franchini M, Bonfanti C. Evolutionary aspects of ABO blood type in humans. China Chimica Acta. 2015;**444**:66-71. DOI: 101016/cca.2015.02.016

Chapter 3

ABO Blood Group and Thromboembolic Diseases

Yetti Hernaningsih

Abstract

Thromboembolic diseases are usually inherited in the family. The tendency to repeat in an individual is a phenomenon that allows it to be studied. The inheritance and recurrence of thromboembolic diseases, of course, have individual risk factors for this occurrence. In the past, the ABO blood group was only needed for transfusion and organ transplant therapy. Over time, scientists think that blood type is a risk factor for certain diseases, including thromboembolism. Many studies divide between type O and non-O blood groups, both of which are distinguished by the presence of antigens on the cell surface and antibodies in the plasma of individuals. Type O does not have A, B antigens but has antibodies against A, B antigens, and vice versa for the non-O type. Many studies have shown that the non-O blood group has a risk factor for thromboembolic diseases, commonly due to higher levels of von Willebrand factor (VWF) and factor VIII (FVIII). These thromboembolic events can occur in arteries or venous. Thromboembolic manifestations are often associated with cardiovascular diseases for arterial thrombosis; and deep vein thrombosis (DVT) and pulmonary embolism (PE) for venous thromboembolism (VTE).

Keywords: ABO, blood group, arteries, venous, thromboembolic diseases

1. Introduction

The known blood types in the population are A, B, AB, and O. These blood types are inherited from both parents. Antigens in the ABO blood group are complex carbohydrates, found in erythrocytes, lymphocytes, platelets, epithelial and endothelial cells, and organs such as the kidneys. Soluble forms of antigen are also synthesized and secreted by tissue cells [1]. The distribution of ABO blood groups in the population depends on race. For example, in India and Mexico, blood type O is the most common. If data from countries in the world are compared, in India and neighboring countries such as Bangladesh and Pakistan, groups O and B dominate, while populations in Europe and Africa are dominated by groups O and A. This comparison explains that the heterogeneity of blood groups in different places and populations is caused by genetic and environmental factors [2].

Initially, the importance of ABO blood type is needed to obtain a match between donor and recipient in the case of transfusion or organ transplantation. Furthermore, many studies reported the association of blood type with a certain disease, especially in the distinction between blood group O and non-O. The non-O blood type has been reported to be associated with several diseases, including cardiovascular disorders, and the incidence of venous thromboembolism (VTE). The first observation on the association between ABO blood type and VTE was made in

IntechOpen

1963 by Dick et al. who found a statistically significant predominance of group A in 461 VTE patients [2–5].

The non-O blood group was at higher risk of thromboembolism due to higher levels of von Willebrand factor (VWF) and factor VIII (FVIII). The rate of proteolytic clearance of VWF by a disintegrin-like and metalloprotease with thrombospondin type 1 motif, 13 (ADAMTS-13) was relatively lower in plasma of the non-O blood group, resulting in a longer VWF half-life in the non-O plasma group than in the O plasma group. As a result, VWF levels were 25–30% higher in non-O group plasma than in group O plasma. High levels of VWF in non-O plasma always lead to increased FVIII levels due to the physiological role of VWF as carriers of FVIII and protecting it from the proteolytic effects of ADAMTS-13. Higher VWF and FVIII levels in subjects with non-O blood groups were strongly correlated with an increased risk of venous thrombosis, a situation that led to non-O blood type being assessed as a genetic risk factor for venous thromboembolism [6].

Wu et al. result in a meta-analysis study of the association between the ABO blood group and vascular disease, the combined odds ratio of the 21 studies analyzing VTE was 1.79 (95% CI, 1.56–2.05) for the non-O versus O group. In three studies in which blood type genotypes were performed, the combination A1 A1/A1 B/BB gave an odds ratio of 2.44 (95% CI 1.79–3.33), while the odds ratio for A1 O/BO/A2 B was 2.11 (95% CI 1.66–2.68), suggesting that the risk is related to the expression of the O(H) antigen [7].

These results are similar to the study of Spiezia et al. in a retrospective case–control study conducted on Italian patients with DVT and controls, which found that non-O blood group increased the risk of DVT by 2.2-fold than individuals with group O. An up to the 7-fold increased risk of VTE was observed when the condition inherited thrombophilic (factor V Leiden, prothrombin G20210A mutations, antithrombin, protein C and protein S deficiency) were associated with non-O blood group carriers compared with non-thrombophilic group-O carriers [4].

The data presented by Spiezia et al. cautioned that the high prevalence of non-O blood type in the general population appears to be one of the most important genetic risk factors for venous thrombosis. Like inherited thrombophilic factors (i.e., factor V Leiden and the prothrombin G20210A mutation), non-O blood types are responsible for a moderate increase in the risk of VTE, and, therefore, ABO blood group testing is recommended in individuals with thrombophilia to assess risk thrombotic [7].

2. ABO blood group

The ABO blood group system was discovered by Landsteiner in 1900. A few years later, von Decastello and Sturly discovered the AB type. Landsteiner's rule stipulates that normal individuals have ABO antibodies against an antigen not found on red blood cells. Individuals with blood type A have A antigens, do not have B antigens, therefore these individuals have B antibodies. On the other hand, individuals with blood group B have B antigens, do not have A antigens, therefore they have A antibodies. The four phenotypes were derived from the two main antigens (A and B) of the system. The phenotypes are group A, group B, group AB, and group O. Individuals with blood type AB means they have A, B antigens and do not have antibodies against A, B antigens. On the other hand, individuals with blood type O have antibodies against antigens A, B and do not have antigens A, B [1]. This classification is important for the sake of blood transfusions that must meet certain requirements.

The A and B alleles are located on chromosome 9 at the ABO locus, encoding the A- and B glycosyltransferase enzymes. ABO antigens can be found on blood cells, lymphocytes, platelets, most epithelial and endothelial cells, and organs such as the kidneys. Soluble forms of antigen can be synthesized and secreted by tissue cells. Soluble antigens can be detected in all body fluids except cerebrospinal fluid. The ABO antigen attached to red blood cells is in the form of glycolipid molecules or glycoproteins, while the main soluble form is the glycoprotein form. Discussion of the ABO antigen requires an understanding of the H antigen. The H gene is located on a different chromosome from the ABO genetic locus and plays a role in controlling the production of H antigen. In addition to the ABO and H genes, the expression of soluble ABO antigens is influenced by the inheritance of the Se gene. The Se gene genetically influences the formation of ABO antigens in saliva, tears, and other body fluids. Consequently, the occurrence and location of ABO antigens are influenced by three genetically independent loci: ABO, H, and Se [1].

The antigen-building block structure for A, B, and H antigens is an oligosaccharide chain attached to a carrier molecule either a protein or lipid. The oligosaccharide chain comprises four sugar molecules linked in simple linear forms or complex branched structures. The two-terminal sugars, d-galactose and N-acetylglucosamine, are coupled in two different configurations. When carbon number 1 of d-galactose is coupled with carbon number 3 of N-acetylglucosamine, the bond is symbolized as $1 \rightarrow 3$. When the number 1 carbon of d-galactose is coupled with the number 4 carbon of N-acetylglucosamine, the bond is described as $\beta 1 \rightarrow 4$. The structure of $\beta 1 \rightarrow 4$ is associated primarily with glycolipids and glycoproteins on the red cell membrane, the structure of $\beta 1 \rightarrow 3$ is located in body fluids and secretions [1].

This transferase catalyzes the addition of certain sugar residues so that the core structure of the H glycan precursor is converted to antigen A (GalNAc 1 → 3 [Fuc 1 → 2] Galβ 1 → 4 GlcNAc 1→), or antigen B (Gal 1–3 [Fuc 1–2] Galβ 1–4 GlcNAc 1-). As a result, the A and B structures are differentiated based on a single terminal sugar (N-acetylgalactosamine versus D-galactose respectively). Individual O groups lack A- or B-transferase activity result from the inactivation of the A1 glycosyltransferase gene, and the nonreducing ends of the corresponding glycans, and therefore continue to express the basic structure of the glycan H (Fuc 1–2] Galβ 1–4 GlcNAc 1-) at the ends of their oligosaccharide chains [8, 9]. Individuals who synthesize determinant A exclusively have blood type A and have genotypes AA or AO, individuals with blood group B are BB or BO, and individuals expressing one allele A and one B have genotype AB. Individuals with blood type O expressing inactive glycosyltransferase A/B have genotype OO. They express only the H antigen. In terms of nomenclature, blood group O includes the H antigen and sometimes the term ABO(H) is used [10].

3. Thromboembolic diseases

Venous and arterial thrombotic disorders have different pathophysiological entities, as a result of anatomical differences and different clinical presentations. In particular, arterial thrombosis results from the phenomenon of platelet activation, whereas venous thrombosis is largely a matter of activation of the clotting system [1].

There are fundamental pathophysiological differences between arterial and venous thrombus. Arterial thrombi which arc happened in small arteries and arterioles are occlusive. Thrombus that occurs in the ventricles of the heart and the great arteries and the aorta, the common carotid artery is nonocclusive. Arterial thrombus is formed in response to increased local shear and exposure to thrombogenic material in damaged vessels, occurs in high-pressure and high-flow systems. Arterial thrombus, referred to as white thrombus, due to consists mainly of platelets and a small amount of fibrin or red blood cells. Leukocytes are also actively recruited to platelet-rich arterial thrombi [11].

The differences from a clinical point of view are as follows: (1) hereditary hypercoagulation (occurs in the "thrombophilia" state), characterized by chronic hyperactivation of the coagulation system, this condition mainly associated with venous rather than arterial thrombosis; and (2) anticoagulant agents (e.g., heparin, warfarin) are usually used to prevent venous thrombosis, whereas antiplatelet agents (e.g., aspirin) are used to prevent arterial thrombosis. In both types of thrombus consists of platelets, fibrin, erythrocytes, and leukocytes with different compositions. Moreover, all thrombi are undergoing propagation, organization, embolization, lysis, and thrombosis, and this dynamic remodeling results in a changing composition [11].

3.1 Arterial thromboembolism

Rudolph Virchow describes three conditions that induce thrombosis, called the Virchow triad. This triad includes endothelial injury, blood flow stasis or turbulence, and blood hypercoagulability. Abnormalities of one or more of these conditions more often manifest the occurrence of DVT. DVT after trauma is more common in conditions of stasis and endothelial injury while spontaneous DVT is more common in cases of hypercoagulability. Risk factors can be classified as acquired or genetic. Genetic risk factors can be divided into strong, moderate, and weak factors. Strong risk factors include deficiency of antithrombin, protein C and protein S. Moderate risk factors include factor V Leiden, prothrombin 20210A, non-O blood type, and fibrinogen 10034C > T. Weak genetic risk factors include fibrinogen variants, factor XIII, and factor XI [11, 12].

Normal wall shear rates range from 300 to 800/s in the large arteries and increase to about 500 to 1600/s in the arterial of microcirculation. However, in pathological stenotic vessels, the wall shear rate can be up to 10,000/s or even higher. The increased shear stress in the microenvironment of the atherosclerotic plaque area of the stenotic vessel is exacerbated by turbulent blood flow. This high hemodynamic force can activate platelets as they pass through the region. This abnormal flow can cause local endothelial dysfunction. High shear stress, especially with a marked shear gradient around the site of the stenosis, is sufficient to induce VWF from endothelial cells and binding of VWF to platelets via glycoprotein Ib-V-IX. This interaction does not occur in normal circulation, result mediating platelet adhesion to the intima surface and triggering platelet thrombus formation [11].

Heterogeneity is seen in the composition of atherothrombotic plaques, even within the same individual. In addition to plaque composition, differences in the basic structural features of the arteries contribute to differences in thrombogenic substrates. For example, the carotid and iliac arteries contain relatively more elastic fibers and proportionately fewer smooth muscle cells than the coronary arteries. Therefore, coronary artery thrombosis usually results in slightly stenotic, lipid-rich plaque, whereas carotid artery usually results in severe stenotic and high-risk plaque [11].

3.2 Venous thromboembolism (VTE)

Venous thrombi are formed mainly from fibrin and red blood cells. Thrombogenic stimulation is caused by (1) stasis of veins, (2) activation of clotting factors, and (3) vascular damage. Anti-thrombogenic properties through mechanisms (1) inactivation of activated coagulation factors by natural inhibitors such as antithrombin and activated protein C, (2) elimination of activated coagulation factors and soluble fibrin polymer complexes by mononuclear phagocytic cells and liver, and (3) lysis of fibrin by enzymes fibrinolytic from plasma and endothelial cells [11].

In the adult group, the predisposition factors to VTE are increasing age, cancer, prolonged immobilization, stroke or paralysis, varicose veins, prolonged air travel, acute inflammatory bowel disease, rheumatic disease, and nephrotic syndrome, oral contraceptive pills, especially those containing third-generation progestins. In the pediatric group, the risk factors for thromboembolism are central venous lines, cancer, and chemotherapy [13].

Venous thrombi are almost always occlusive and can form casts of the vessels in which they arise. Unlike arterial thrombus, severe vascular damage is generally not found at the site of venous thrombosis. Therefore, in low-flow and low-pressure venous systems, decreased blood flow (stasis) and systemic activation of the coagulation cascade play a major pathophysiological role. Venous thrombi consist mostly of red blood cells trapped in fibrin and contain relatively few platelets; hence, they have been described pathologically as red thrombi [11].

The study of Sun et al. in 1412 patients with VTE (consisting of 600 DVT patients, 441 PE patients, and 371 patients having a diagnosis of DVT and PE) and 199,248 controls the results of VTE patients were significantly higher in the non-O blood group compared to all non-VTE discharge patients with OR 1.362 (95% confidence interval, 1.205-1.540). When the non-O group was classified into A, B, and AB and a pairwise comparison test was performed on VTE and non-VTE patients, the results were not statistically different [14].

4. The relation of ABO blood group and thromboembolic diseases

ABO blood group has been recognized as a risk factor for thromboembolic diseases since the 1960s. Many studies have shown that the non-O group had a higher incidence of ischemic heart disease. ABO blood type is important in relation to VWF and FVII levels because in turn confer a clear risk of increased VTE especially in non-O blood groups which provide a higher increase. This association is less clear for CAD and MI but a similar pattern emerges with most studies finding group O to be at lower risk [15].

The Framingham Heart Study, and others, resulted in blood group A having an increased risk of CAD [16–18] and MI [19]. More specifically, blood type A is associated with early detection of CAD [19, 20] and predominates in patients with MI [21]. Another study reported that groups B [22, 23] or AB [24] had a higher incidence of CAD. In contrast, Mitchell [25] reported that cities with a higher prevalence of group O had higher rates of cardiovascular mortality and a study in India showed that blood type O increased the risk of CAD [26]. Further studies did not identify any association between blood type and CAD [27, 28]. Based on these inconsistent results, He et al. [29] conducted a meta-analysis found the highest risk of CAD was observed in blood group AB, followed by groups B, A, and O. This is similar to ABO-associated vWF/FVII levels which the highest in group AB, followed by groups B, A and O [18].

The theory proposed to explain the relationship between ABO blood group and CAD is as follows. Fibrinogen together with vWF activates platelet aggregation and adhesion which in turn plays a role in the development of atherosclerosis. On the other hand, blood group A has been reported to have higher cholesterol levels and lower lipoprotein density, this may explain the association with an increased risk of CAD. In addition, ABO loci have been reported to be associated with inflammatory-forming CAD, including intercellular adhesion molecule-1, soluble P selectin, soluble E selectin, and tumor necrosis factor-α. Meanwhile, the interaction between genetic factors (genes known to increase susceptibility to CAD and the ABO locus) and environmental factors still contribute to the risk of CAD and MI [15].

The incidence of VTE is more often due to factor VIII (FVIII) and von Willebrand Factor (VWF) levels are higher in the non-O blood group than in the O blood group. Moeller et al. comparing VWF and FVII levels in individuals with ABO phenotype found the following order O < A < B < AB for vWF levels and O < A < AB < B for FVII levels [30]. Nevertheless, Simangunsong et al. found no significant differences were present in factor VIII activity between A, B, and O blood types [31].

A blood type that is identical to high vWF, is an important genetic factor that explains around 30% of the variation in factor VIII levels. There is a relationship between factor VIII and vWF. However, attempts to find other genetic loci associated with high vWF and factor VIII levels have not been successful to date. Most likely, the high factor VIII levels are due to increased synthesis or decreased clearance of the vWF-factor VIII complex [32]. Furthermore, non-O blood groups are associated with increased arterial and venous thrombotic events possibly mediated by increased levels of von Willebrand factor and factor VIII in non-O blood groups [33].

Meanwhile, several studies have confirmed that the level of vWF is lower in people who have blood type O, therefore FVIII: C is reduced due to the stabilizing function of vWF as an FVIII: C carrier. Factor VIII affinity for vWF may also differ from individual to individual, which is genetically determined [30].

Blood group A is associated with an increased odds of major adverse cardiovascular events (MACE), whereas blood group O was associated with a reduction in the odds of MACE in patients with COVID19. These findings suggest an association between blood group type and cardiovascular complications in COVID-19. The biological mechanism behind the role of ABO blood groups in COVID-19 remains elusive. Natural anti-glycan ABO antibodies have been shown to inhibit SARS-CoV1 interaction of spike protein and angiotensin-converting enzyme 2 (ACE2) [33].

In the cellular experimental model approach, it can be proven that the binding of the SARS-CoV S protein with ACE2 on target cells can be blocked by anti-A antibodies in the blood group, because the S protein is synthesized by A-antigen-expressing cells, after transfection by cDNA glycosyltransferases. in accordance. When produced in cells expressing blood type A or B enzymes, SARS virions are decorated by appropriate glycan antigens, consequently, the presence of anti-A and anti-B antibodies in blood type O individuals can block the attachment and entry of the virus thereby preventing infection. Therefore, individuals with blood type O will have a lower risk of infection than non-O individuals. This phenomenon occurred during the 2003 Hong Kong SARS hospital outbreak, and a similar trend was recently observed for COVID-19 in China, the infectious SARS virions are decorated by glycan antigens corresponding to blood group A or B, and the presence of anti-A antibodies and anti-B in individuals with blood type O can prevent infection by blocking the attachment and entry of the virus [34, 35].

Vasan study used data on 1.1 million healthy blood donors from the binational database SCANDAT2 (Scandinavian Donations and Transfusions), which contains national data on blood donation and transfusion from Sweden and Denmark, to investigate the association between ABO blood type and arterial thrombotic events. or veins. And the results confirm that there is a consistent relationship between non-O blood type and VTE and cardiovascular events, with a greater risk in the venous. The proposed basic mechanisms driving this association include higher concentrations of factor VIII and von Willebrand factor in individuals with non-O blood types. This study provides strong evidence of a consistent relationship between the non-O blood group and VTE, and the incidence of cardiovascular thrombosis, with a greater risk of recurrence in non-O blood groups. Also, non-O blood groups confer an increased risk of thromboembolism, ABO blood groups may have a role in thrombosis risk assessment and could potentially be added to existing clinical prediction systems [5].

Sickle cell disease or what is known as sickle cell trait (SCT) in individuals will provide a risk of DVT even though it is weak. This potential risk will increase if the patient has a non-O blood group. This combined effect will increase the activation of clotting factors and increase the risk of DVT. Thus, caution should be exercised in co-inheritance of non-O blood group and SCT, in which case it should be paid attention to assess the risk profile of DVT in patients in Africa and other areas where SCT is common [6].

Non-group O patients have susceptibility and greater risk of VTE than patients of group O and have greater levels of von Willebrand factor (vWF) and factor VIII. The risk of VTE is probably related to the level of vWF and factor VIII. A, B, and H blood group antigens are expressed on N-glycans of VWF and influence the half-life of the protein (10 hours for group O and 25 hours for non-O subjects), explaining the greater levels in non-O patients [8].

In the report of Rejtő et al. who investigated the effect of ABO, VWF level, age on the variability of F VIII levels in 8 patients non-severe Hemophilia A results that ABO and VWF levels did not influence the variability of FVIII levels, whereas age had only a small influenced [36].

The coagulation process is under the control of several inhibitors which limit clot formation. A balance between procoagulants and anticoagulants is necessary for maintaining hemostasis. A thrombus is formed as a result of a disturbance in this balance. Thrombus is formed when the procoagulant activity of one of the coagulation factors is increased or the activity of one of the natural inhibitors is decreased, a condition called thrombophilia can occur in inherited deficiency of natural inhibitors, as well as with inherited gain-of-function mutations of some coagulation factors. The deficiency of natural inhibitors such as antithrombin, protein C and inherited protein S is a strong risk factor for venous thrombosis; they have little or no effect on arterial thrombosis. Antithrombin directly inhibits several activated coagulation factors, notably thrombin, and activated factor X, and the inhibitory effect is amplified by its binding to glycosaminoglycans on the endothelial surface carrying heparin-like activity. The effect of increasing the tendency for clot formation is especially in the venous system where the coagulation pathway (different from that of platelets) plays a major role. The anticoagulant protein C on the surface of the endothelium is very important in the down-regulation of thrombin formation. Activated protein C inactivates factor Va and factor VIIIa proteolytically, the two most important activated cofactors of the coagulation cascade, causing a slowdown in the rates of thrombin and fibrin formation. The inhibitory effect of activated protein C is accelerated by its main cofactor, protein S. Inherited deficiency of one of these inhibitors leads to increased thrombin formation, increasing susceptibility to VTE [37].

Ahmed et al. hypothesized that if Sickle Cell Trait was a risk factor for DVT, individuals with non-O blood group and SCT (Hb AS) would have a higher risk of DVT than those with non-O blood group and normal hemoglobin (Hb AA) phenotype. The results of this study indicate that SCT itself is a weak risk factor for DVT, but would have the potential to increase the risk of DVT in patients with non-O blood groups. Therefore, co-inheritance of SCT and non-O blood groups is an important risk factor for DVT [6].

The study of Vasan et al. results almost in all age groups, the incidence of VTEs and cardiovascular events is higher in non-O than O blood groups. The incidence rate ratio (IRR) was highest for the venous events, with all venous thrombotic events combined for individuals with non-O blood group compared with blood group O having an IRR of 1.80 (95% CI, 1.71–1.88). The risk patterns were similar for pulmonary embolism and deep vein thrombosis. Among arterial events, IRRs were generally lower with IRRs of 1.10 (95% CI, 1.05–1.14) for myocardial infarction and 1.07 (95% CI, 1.02–1.12) for stroke in individuals in non-O blood groups compared with those in blood group O [5].

5. Summary

ABO blood type is associated with the risk of thromboembolic diseases. Non-O blood type has a greater risk than O blood type. Thromboembolic events occur in both arteries and venous, which are venous more often, one of the causes is FVIII and VWF clearance in non-O blood groups are longer, results found high levels of both in the non-O blood group. While the manifestation of arterial thromboembolism commonly happened in cardiovascular diseases including coronary arterial disease and myocardial infarct.

Conflict of interest

No conflict of interest in this article.

Abbreviations

ADAMTS-13	a disintegrin-like and metalloprotease with thrombospondin type 1 motif, 13
CAD	coronary artery disease
CI	confidence interval
COVID-19	Coronavirus disease-19
DVT	deep vein thrombosis
FVIII	factor VIII
Hb	hemoglobin
IRR	incidence rate ratio
MACE	major adverse cardiovascular events
MI	myocardial infarction
PAR	population-attributable risk percent
PE	pulmonary embolism
PY	patient-year
SCT	sickle cell trait
VTE	venous thromboembolism C1
VWF	von Willebrand Factor

Author details

Yetti Hernaningsih
Faculty of Medicine Universitas Airlangga, Dr. Soetomo General Academic Hospital, Surabaya, Indonesia

*Address all correspondence to: yetti-h@fk.unair.ac.id

References

[1] Howard PR. Basic & Applied Concepts of Blood Banking and Transfusion Practices. 4th ed. Missouri: Elsevier Inc.; 2017. pp. 97-206

[2] Agrawal A, Tiwari AK, Mehta N, Bhattacharya P, Wankhede R, Tulsiani S, et al. ABO, and Rh (D) group distribution and gene frequency; the first multicentric study in India. Asian Journal of Transfusion Science. 2014;**8**(2):121-125. DOI: 10.4103/0973-6247.137452

[3] Canizalez-Román A, Campos-Romero A, Castro-Sánchez JA, López-Martínez MA, Andrade-Muñoz FJ, Cruz-Zamudio CK, et al. Blood groups distribution and gene diversity of the ABO and Rh (D) loci in the Mexican population. BioMed Research International. 2018;**2018**:1925619. DOI: 10.1155/2018/1925619

[4] Franchini M, Mannucci PM. ABO blood group and thrombotic vascular disease. Thrombosis and Haemostasis. 2014;**112**(6):1103-1109. DOI: 10.1160/TH14-05-0457

[5] Vasan SK, Rostgaard K, Majeed A, Ullum H, Titlestad KE, Pedersen OBV, et al. ABO blood group and risk of thromboembolic and arterial disease. A study of 1.5 million blood donors. Circulation. 2016;**133**:1449-1457. DOI: 10.1161/CIRCULATIONAHA.115.017563

[6] Ahmed SG, Kagu MB, Ibrahim UA, Bukar AA. Impact of sickle cell trait on the thrombotic risk associated with non-O blood groups in northern Nigeria. Blood Transfusion. 2015;**13**:639-643. DOI: 10.2450/2015.0335-14

[7] Franchini M, Makris M. Non-O blood group: An important genetic risk factor for venous thromboembolism. Blood Transfusion. 2013;**11**:164-165. DOI: 10.2450/2012.0087-12

[8] Anstee DJ. The relationship between blood groups and disease. Blood. 2010;**115**(23):4635-4643. DOI: 10.1182/blood-2010-01-261859

[9] Ward S, O'Sullivan J, O'Donnell JS. The relationship between ABO blood group, von Willebrand factor, and primary hemostasis. Blood. 2020;**136**(25):2864-2874. DOI: 10.1182/blood.2020005843

[10] Stanley P, Cummings RD. Structures common to different Glycans. In: Varki A, Cummings RD, Esko JD, et al., editors. Essentials of Glycobiology. 3rd ed. Cold Spring Harbor (NY): Cold Spring Harbor Laboratory Press; 2017

[11] Mohler ER III, Schafer AI. Atherothrombosis: Disease initiation, progression, and treatment. In: Kaushansky K, Levi M, editors. Williams Hematology Hemostasis and Thrombosis. Toronto: McGraw-Hill Education; 2018. pp. 467-490

[12] Prandoni P. Venous and arterial thrombosis: Two aspects of the same disease? Clinical Epidemiology. 2009;**1**:1-6. DOI: 10.1016/j.ejim.2009.02.001

[13] Kesieme E, Kesieme C, Jebbin N, Irekpita E, Dongo A. Deep vein thrombosis: A clinical review. Journal of Blood Medicine. 2011;**2**:59-69. DOI: 10.2147/JBM.S19009

[14] Sun X, Feng J, Wu W, Peng M, Shi J. ABO blood types associated with the risk of venous thromboembolism in Han Chinese people: A hospital-based study of 200,000 patients. Scientific Reports. 2017;7:42925. DOI: 10.1038/srep42925

[15] Zhou S, Welsby I. Is ABO blood group truly a risk factor for thrombosis and adverse outcomes? World Journal of Cardiology. 2014;**6**(9):985-992. DOI: 10.4330/wjc.v6.i9.98

[16] Garrison RJ, Havlik RJ, Harris RB, Feinleib M, Kannel WB, Padgett SJ. ABO blood group and cardiovascular disease: The Framingham study. Atherosclerosis. 1976;**25**:311-318. DOI: 10.1016/0021-9150(76)90036-8

[17] Whincup PH, Cook DG, Phillips AN, Shaper AG. ABO blood group and ischaemic heart disease in British men. BMJ. 1990;**300**:1679-1682. DOI: 10.1136/bmj.300.6741.1679

[18] Rosenberg L, Miller DR, Kaufman DW, Helmrich SP, Van de Carr S, Stolley PD, et al. Myocardial infarction in women under 50 years of age. JAMA. 1983;**250**:2801-2806. DOI: 10.1001/jama.1983.03340200035025

[19] Lee HF, Lin YC, Lin CP, Wang CL, Chang CJ, Hsu LA. Association of blood group a with coronary artery disease in young adults in Taiwan. Internal Medicine. 2012;**51**:1815-1820. DOI: 10.2169/internal medicine. 51.7173

[20] Sari I, Ozer O, Davutoglu V, Gorgulu S, Eren M, Aksoy M. ABO blood group distribution and major cardiovascular risk factors in patients with acute myocardial infarction. Blood Coagulation & Fibrinolysis. 2008;**19**: 231-234. DOI: 10.1097/MBC. 0b013e3282f54522

[21] Carpeggiani C, Coceani M, Landi P, Michelassi C, L'abbate A. ABO blood group alleles: A risk factor for coronary artery disease. An angiographic study. Atherosclerosis. 2010;**211**:461-466. DOI: 10.1016/j.atherosclerosis.2010. 03.012

[22] Nydegger UE, Wuillemin WA, Julmy F, Meyer BJ, Carrel TP. Association of ABO histo-blood group B allele with myocardial infarction. European Journal of Immunogenetics. 2003;**30**:201-206. DOI: 10.1046/j. 1365-2370.2003.00390.x

[23] Stakisaitis D, Maksvytis A, Benetis R, Viikmaa M. Coronary atherosclerosis and blood groups of ABO system in women (own data and review). Medicina (Kaunas, Lithuania). 2002;**38**(Suppl. 2):230-235

[24] Meade TW, Cooper JA, Stirling Y, Howarth DJ, Ruddock V, Miller GJ. Factor VIII, ABO blood group and the incidence of ischaemic heart disease. British Journal of Haematology. 1994;**88**:601-607. DOI: 10.1111/j.1365-2141.1994.tb05079.x

[25] Mitchell JR. An association between abo blood-group distribution and geographical differences in death-rates. Lancet. 1977;**1**:295-297

[26] Biswas S, Ghoshal PK, Halder B, Mandal N. Distribution of ABO blood group and major cardiovascular risk factors with coronary heart disease. BioMed Research International. 2013;**2013**:782941. DOI: 10.1155/ 2013/782941

[27] Amirzadegan A, Salarifar M, Sadeghian S, Davoodi G, Darabian C, Goodarzynejad H. Correlation between ABO blood groups, major risk factors, and coronary artery disease. International Journal of Cardiology. 2006;**110**:256-258. DOI: 10.1016/j. ijcard.2005.06.058

[28] Biancari F, Satta J, Pokela R, Juvonen T. ABO blood group distribution and severity of coronary artery disease among patients undergoing coronary artery bypass surgery in northern Finland. Thrombosis Research. 2002;**108**: 195-196. DOI: 10.1016/S0049-3848(03)00003-3

[29] He M, Wolpin B, Rexrode K, Manson JE, Rimm E, Hu FB, et al. ABO blood group and risk of coronary heart disease in two prospective cohort studies. Arteriosclerosis, Thrombosis, and Vascular Biology. 2012;**32**:2314-2320. DOI: 10.1161/atvbaha.112.248757

[30] Moeller A, Weippert-Kretschmer M, Prinz H, Kretschmer V. Influence of ABO blood groups on primary hemostasis. Transfusion. 2001;**41**:56-60. DOI: 10.1046/j.1537-2995.2001.41010056.x

[31] Simangunsong AD, Hernaningsih Y. Comparison of factor Viii activity In O and non-O blood types (Perbandingan Aktivitas Faktor VIII antara Golongan Darah O dan non-O). Indonesian Journal of Clinical Pathology and Medical Laboratory. 2017;**23**(3):221-224. DOI: 10.24293/ijcpml.v23i3.1197

[32] Kamphuisen PW, Eikenboom JCJ, Bertina RM. Elevated factor VIII levels and the risk of thrombosis. Arteriosclerosis, Thrombosis, and Vascular Biology. 2001;**21**:731-738. DOI: 10.1161/01.ATV.21.5.731

[33] Nauffal V, Achanta A, Goldhaber SZ, Piazza G. Association of ABO blood group type with cardiovascular events in COVID-19. Journal of Thrombosis and Thrombolysis. 2021;**51**:584-586. DOI: 10.1007/s11239-020-02364-5 1 3

[34] Breiman A, Ruvën-Clouet N, Le Pendu J. Harnessing the natural anti-glycan immune response to limit the transmission of enveloped viruses such as SARS-CoV-2. PLoS Pathogens. 2020;**16**(5):e1008556. DOI: 10.1371/journal

[35] Guillon P, Clément M, Sébille V, Rivain JG, Chou CF, Ruvoën-Clouet N, et al. Inhibition of the interaction between the SARS-CoV spike protein and its cellular receptor by anti-histo-blood group antibodies. Glycobiology. 2008;**18**(12):1085-1093. DOI: 10.1093/glycob/cwn093 Epub 2008 Sep 25

[36] Rejtő J, Königsbrügge O, Grilz E, Hofer S, Mauracher L-M, Gabler C, et al. Influence of blood group, von Willebrand factor levels, and age on factor VIII levels in non-severe haemophilia a. Thrombosis and Haemostasis. 2020;**18**:1081-1086. DOI: 10.1111/jth.14770

[37] Previtali E, Bucciarelli P, Passamonti SM, Martinelli I. Risk factors for venous and arterial thrombosis. Blood Transfusion. 2011;**9**(2):120-138. DOI: 10.2450/2010.0066-10

Chapter 4

Duffy Antigens and Malaria: The African Experience

Chima Akunwata

Abstract

The Duffy blood group antigen is also known as Duffy Antigen Receptor for Chemokines (DARC) serves more functions than just a blood group antigen for serological reactions. It is a receptor for pro-inflammatory chemokines and *Plasmodium vivax* invasion of the red blood cells. A point mutation in the promoter region of the Duffy gene disrupts the binding of a transcription factor, leading to a lack of expression of the antigen on the erythrocytes. This Duffy negative phenotype is found predominantly in the African population. This mutation is advantageous as individuals with the Fy(a-b-) phenotype are less susceptible to *P. vivax* malaria. Malaria is caused by plasmodium parasites and it is endemic in Africa, where it is one of the leading causes of morbidity and mortality. It is believed that the absence of Duffy antigen in most Africans contributed to the resistance to *P. vivax* and by extension, reduced the burden of malaria in these endemic areas.

Keywords: Duffy antigen, DARC, Duffy-negative, *P. vivax*, malaria, Africa

1. Introduction

The red blood cell membranes have numerous antigenic determinants—carbohydrates, proteins, or the combination of the two, the knowledge of which has been employed in immunohematology in the provision of safe blood. These inherited however, may subserve other physiological functions or be involved in many pathological conditions or disease susceptibility [1, 2].

Duffy blood group system is important in clinical medicine where it may be involved in transfusion reaction, hemolytic disease of the fetus and newborn (HDFN), and as chemokine receptors. More importantly, it determines susceptibility to *Plasmodium vivax* infection. The absence of the antigens in red blood cells of Africans provides an apparent explanation for the protection against the parasite [3].

Malaria is endemic in the tropics, but the susceptibilities to infection are not the same. The African populations are largely resistant to *P. vivax* and *P. knowlesi* infections due to the absence of the Duffy antigen, a critical receptor in the invasion of human host red cells by these species of malaria parasite [4, 5].

This chapter serves as a synopsis of the Duffy antigen and malaria infection. It presented the history of the discovery of Duffy antigens, their gene, protein, antibodies, and functions. The selective pressure of malaria on the Duffy gene and mutation that led to a protective phenotype against the *P. vivax* malaria infection in Africa were discussed.

2. Duffy blood group system

Duffy blood group was first described in 1950 by Cutbush in a multiply transfused hemophilic patient [6, 7]. The alloantibody against the antigen was designated as Fy^a. The antibody was named after the patient. A year later, another antibody was described in the serum of a multiparous woman and was designated anti-Fyb. Duffy antigen maps to the long arm of chromosome 1 at position 1q22 → q23 while the RH gene is the short arm of the same chromosome [8]. The significance of the co-location of these genes on the same chromosome is that their interaction produced the Fy5 antigen as red cell from individuals with Fy(a-b-) and the Rh$_{null}$ phenotype lack this antigen [9].

2.1 Duffy antigen

Duffy antigen is a glycoprotein. It is also known as the Duffy antigen receptor for chemokines (DARC). It is a seven-transmembrane helix receptor with the N-terminus in the extracellular domain while the C-terminus forms the intracellular domain (**Figure 1**) [10]. It has a structural similarity with G-protein-coupled receptor but is not a member of this family [11, 12]. There are six known Duffy antigens—Fy^a, Fy^b, Fy^3, Fy^4, Fy^5, Fy^6 and four phenotypic expressions—Fy(a+b+), Fy(a-b+), Fy(a+b-), and Fy(a-b-) as shown in **Tables 1** and **2** respectively. The most common antigens are Fy^a and Fy^b. The Fy^x antigen results from the weak expression of Fy^b, is found in whites and is due to a single mutation in the FYB gene. The Duffy null phenotype Fy(a-b-) occurs in about two-thirds of the black population, while it is rare in Caucasians. The genetic basis of this null phenotype is distinct in these populations (see genetic basis below).

Duffy antigen is a receptor for chemokines in the C-X-C class (e.g., interleukin-8 (IL-8) and C-C class (e.g., MCP-T). The physiological function of this receptor is

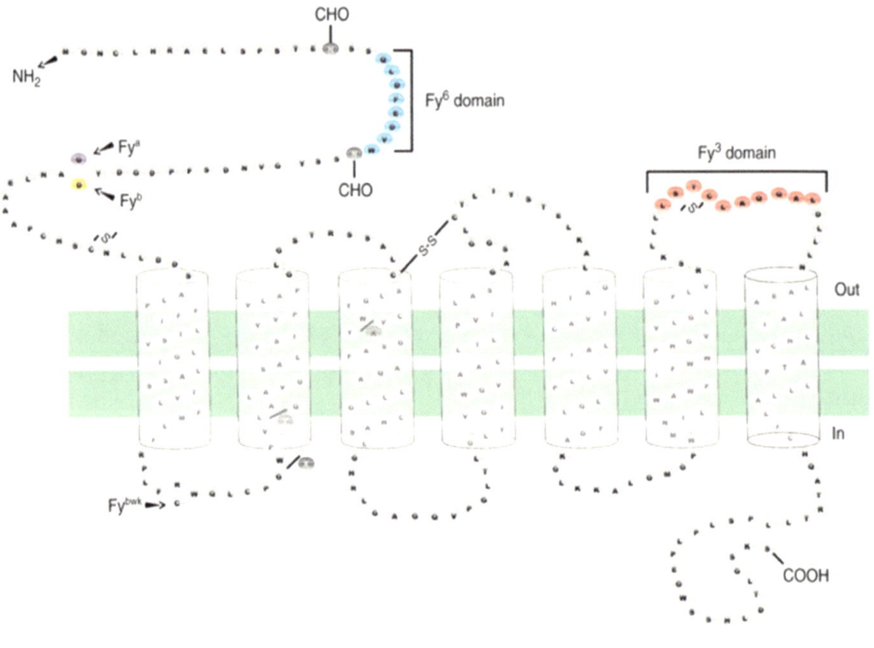

Figure 1.
Duffy glycoprotein seven-transmembrane domain structure. Amino acid changes responsible for the Fya/Fyb polymorphism, the Fy^x mutation, and Fy3 and Fy6 regions [10].

Antigen	ISBT symbol	ISBT number
Fya	FY1	008001
Fyb	FY2	008002
Fy3	FY3	008003
Fy4	FY4	008004
Fy5	FY5	008005
Fy6	FY6	008006

ISBT—International Society of Blood Transfusion.

Table 1.
The Duffy antigens and ISBT symbols and numbers.

Prevalence %			
Red cell phenotype	**Caucasians**	**Blacks**	**Alleles**
Fy(a+b-)	20	10	FY*01/FY*01 or FY*A/FY*A
Fy(a-b+)	32	20	FY*02/FY*02 or FY*B/FY*B
Fy(a+b+)	48	3	FY*A/FY*B
Fy(a-b-)	Rare	67	FY*/N.01–05, FY*/N.01–02
Fy^3	100	32	
Fy^5	99.9	32	
Fy^6	100	32	

Table 2.
Duffy phenotypes, prevalence and alleles.

to modulate the blood–tissue gradient of these cytokines during immune responses [13]. The red blood cells through the DARC receptors act as adsorption surfaces or as chemokine scavengers for inflammatory cytokines such as IL-8, thereby eliminating excess chemokines during immune responses. Duffy antigens are express on epithelial cells of capillary and post-capillary venules, epithelial cells of the kidney collecting ducts, lung alveoli, and Purkinje, cells of the cerebellum.

2.2 Duffy antibodies

The Duffy antibodies are rarely naturally occurring. Anti-Fy^a and anti-Fy^b are IgG antibodies in the IgG1 subclass [14]. They result from sensitization after transfusion or pregnancy. Anti-Fy^a is more frequently encountered than Anti-Fy^b. In the black population with Fy(a-b-) phenotype anti-Fy^a is produced but not anti-Fy^b. Contrastingly, white individuals with rare Fy(a-b-) produce anti-Fy^3. Anti-Fy^3, -Fy^4, -Fy^5, have been described but no human anti-Fy^6 has been identified but a mouse monoclonal antibody has been raised against Fy^6 epitope.

3. Genetic basis and biochemistry of Duffy antigen

The gene, *ACKR1* also known as *DARC* or *FY*, that encodes the Duffy blood group antigens is located at chromosome 1q23.2. The two allelic forms *FYA* and *FYB* differ by a single nucleotide at position c.125G>A and define the Fy(a+b-), Fy(a-b+), and Fy(a+b+). The gene products differ by a single amino acid at residue

42—glycine and aspartic acid respectively [3]. The Fy(a-b-) phenotypes (Duffy negative) seen in many Africans, African Americans, and some European and Asians result from two genetic mechanisms. The most common mutation occurs in the promoter region of the *FYB* allele, where a point mutation c.1-67T>C prevents expression of the antigen on the red blood cells but allows expression on other tissues. This is an erythroid-specific mutation and it is commoner in Africans. A similar mutation has been found *FYA* allele but, it is rarer [8].

In Europeans and Asians, Fy(a-b-) phenotype arises from a mutation in the coding region (a point mutation introduces a premature stop codon) of the *FYA* or *FYB* allele preventing the antigen expression in all tissues. These are true Duffy null phenotypes.

4. Malaria

Malaria remains a major public health problem in tropical and subtropical areas of the world. It is one of the major causes of childhood mortality and an indirect cause of maternal mortality [15].

It is a mosquito-borne disease. The parasite responsible for malaria belongs to the genus *Plasmodium*. The most common species causing human infections include *P. falciparum* which causes malignant tertian malaria, *P. vivax*, benign tertian malaria, *P. ovale* benign tertian, *P. malariae* benign quartan and *P. knowlesi* quotidian malaria. The lifecycle of the *Plasmodium* spp. is complex with the sexual phase occurring in the mosquito vector (*Anopheles* genus) and the asexual phase in the human host [16]. Infected female *Anopheles* mosquitoes inject sporozoites into the human host during a blood meal. The sporozoites gain access to the hepatocytes within 30–60 min where they form merozoites through asexual reproduction. These merozoites are released into the bloodstream where they parasitize and replicate with the red blood cells (erythrocytic schizogony). Some merozoites differentiate into male and female gametocytes that are ingested by the mosquitoes in the next blood meal. The gametocytes produce sporozoites to continue the cycle [17].

The severity of *Plasmodium* infection depends on the species and the host immunity which is the function of previous exposures [17]. *P. vivax* infection usually causes uncomplicated malaria, although severe forms have been reported while *P. falciparum* causes severe malaria. Malaria infection usually presents with fever, abdominal discomfort, headache, joint aches, muscle aches, abdominal discomfort, vomiting, lethargy, anorexia [18]. One of the defining characteristics of *P. vivax* infection is the dormant liver stage (hypnozoites) it forms which reactivates weeks to months after initial infection [19]. Malaria infection is associated with complications such as splenomegaly, thrombocytopenia, derangements in liver enzymes such as raised alanine aminotransferase (ALT), jaundice, renal failure, ARDS, and cerebral malaria [20]. Although *P. vivax* infection is generally benign, it could have these complications similar to *P. falciparum* malaria.

4.1 Epidemiology

Plasmodium falciparum and *P. vivax* are the most common causes of malaria accounting for an estimated 229 million cases in 2019 in 87 endemic countries. Of these cases, 97% are found in Africa especially in sub-Saharan (SSA), and estimated malaria-related deaths of 409,000. Most of these deaths are in Africa, with children aged under 5 being disproportionately affected [15]. The most prevalent Plasmodium parasite outside Africa is *P. vivax* responsible for about 6 million cases [21, 22]. *P. vivax* is the most widely distributed plasmodium species putting over

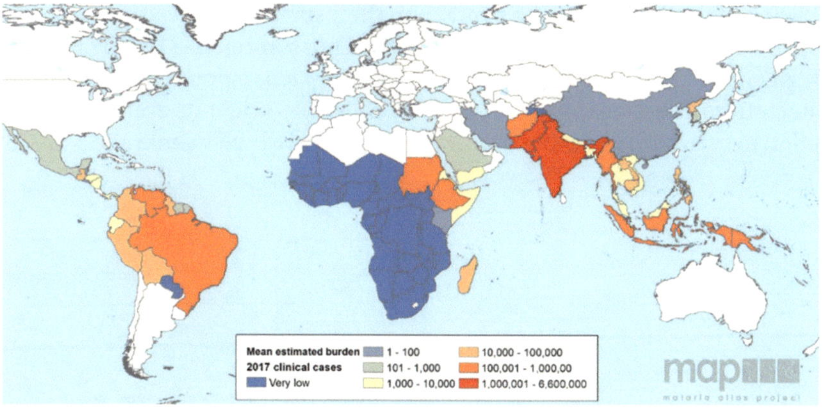

Figure 2.
The mean estimated clinical burdens of P. vivax malaria. Shades of blue color show very low clinical burden to red with high clinical burden [23].

4 billion people at risk of infection. Transmission has been reported in the Horn of Africa, Central and South American, Asia, and Pacific Islands [19]. The largest burden of *P. vivax* malaria occurs in the Indian subcontinent and the horn of Africa. The sub-Saharan Africa has a very low prevalence (**Figure 2**) [23].

Due to the high prevalence of Duffy negative phenotype in the sub-Saharan African, (SSA) there is a relative absence of *P. vivax* malaria infection. However, evidence is accumulating that there are *P. vivax* malaria infections in SSA occurring at lower prevalence [21, 24]. A prevalence of 2.9% was found in a nationwide survey in the Democratic Republic of Congo [21], a seroprevalence of 15.2% was found in Beninese blood donors [25].

4.2 Malaria adaptation and selective pressure

Malaria is known to be a major driving force in evolutionary selection in the human genome [26, 27]. The ethnic differences in susceptibility to malaria infection, the protective effects of G6PD deficiency, thalassemia, and hemoglobin C on severe malaria infection have been linked to this selective pressure. Malaria may also modulate genes involved in immunity inflammation, cell adhesion [26]. There is a strong correlation between the prevalence of negative FYA and FYB alleles, consequently the absence of the Duffy antigens on the endemicity for *P. vivax*. The Duffy negative phenotype was found to be due to a single nucleotide polymorphism in the promoter region leading to disruption of the binding site for GATA-1 erythroid transcription factor and resistance to *P. vivax* invasion of erythrocytes [28]. GATA-1 is one of the nuclear transcription factors in the GATA hemopoietic subfamily. It contributes to erythroid commitment and differentiation [29]. GATA-1 recognizes and binds to GATA consensus binding motif on the Duffy gene. As demonstrated by Tournamille et al., a point mutation on the DARC promoter (CTTA**T**CT → CTTA**C**CT) affects the interaction with the transcription factor [28]. This single nucleotide change from T to C found only in the Duffy-negative genome abolishes or disrupts the binding of GATA-1 to the DARC promoter leading to the absence of Duffy antigen on their red cell membranes [30].

4.3 Mechanism of *P. vivax* invasion of red blood cells

Malaria parasites exhibit different red cell tropisms. *P. vivax* merozoites preferentially bind to reticulocytes than normocytes. The invasion of red blood cells by *P. vivax* depends on the membrane glycoprotein of the Duffy blood group system.

The *P. vivax* merozoites express a protein on their surface, *P. vivax* binding protein (PvDBP) through which they interact with the Duffy antigen [31]. The PvDBP is a 140KD protein with a 330-amino acid cysteine-rich region responsible for this interaction. The merozoite of *P. vivax* is able to re-orient its apical surface in apposition to both Duffy-positive and Duffy-negative red cell membranes [32, 33].

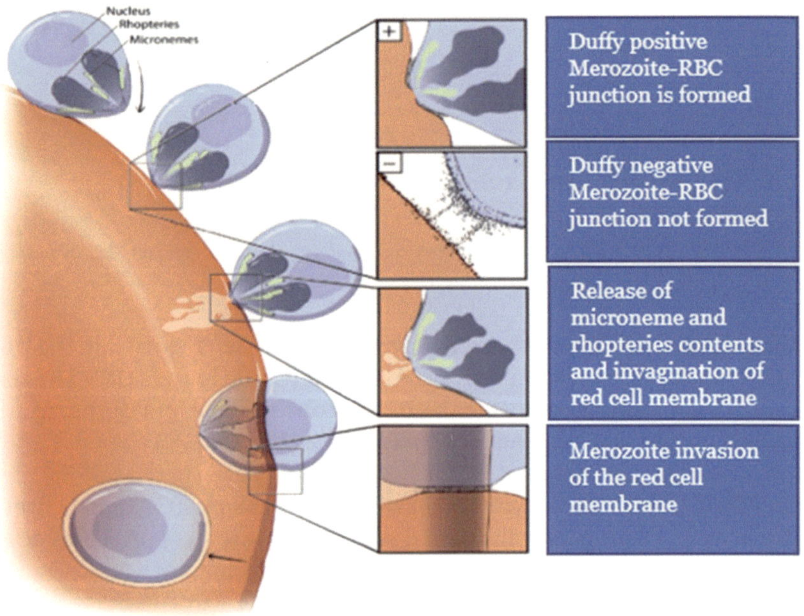

Figure 3.
Overview of P. vivax merozoite interaction with the human red blood cell. Red blood cells without the Duffy antigen are resistant to invasion by P. vivax [31].

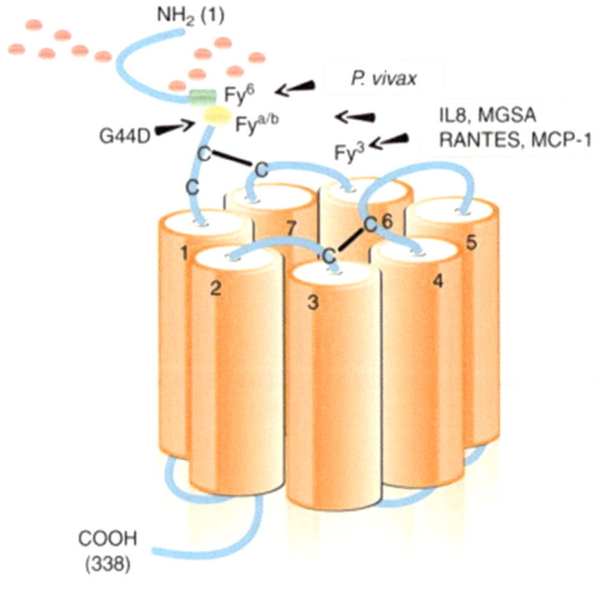

Figure 4.
Duffy glycoprotein showing different interaction sites for the P. vivax and the chemokines [11].

However, the tight junction is not formed between the merozoite and the Duffy-negative red blood cells, suggesting that the Duffy antigen is necessary for the invasion of the red blood cells by the parasite (**Figure 3**) [31].

Consequently, Duffy-negative erythrocytes do not bind to *P. vivax* merozoites [34]. Unlike *P. falciparum* which uses a series of receptors to invade the red cells, *P. vivax* requires the antigens of the Duffy blood system to invade the red blood cells (**Figure 4**) [11]. Thus, in African populations where most have the Fy(a-b-) phenotype, invasion is uncommon. Recently, however, invasion of red cells has been reported in Duffy negative individuals, this suggests that there may be other targets used by the parasite [35]. Susceptibility to *P. vivax* infection has also been shown to exhibit a dosage effect. This means that there are twice as many Fy^a antigens on RBCs from an individual who is homozygous for the Fya allele than on RBCs from an individual who is heterozygous. Consequently, in some populations, carriers of the Fy(a-b+) or Fy(a+b-) have half of the Duffy antigen and reduced ability for their red blood cells to be infected by *P. vivax* [36]. The implication of *P. vivax* preference of parasitizing reticulocytes is that in African populations, where sickle cell anemia and glucose-6-phosphate dehydrogenase deficiency are endemic, and reticulocytosis is a common finding in these disorders due to recurrent hemolytic anemia, a *P. vivax* infection would have been added burden to the mortality and morbidity already caused by *P. falciparum*.

5. Conclusion

Earlier discoveries of the red cell antigens and their antibodies helped provide safe blood for transfusion. In addition to its roles in transfusion medicine, Duffy antigen acts as a receptor for the *P. vivax* malaria parasite and as a receptor for chemokines. The fortuitous mutation that resulted in less susceptibility to the parasite in the African population led to relieving the burdens that would have resulted in synergistic infection with *P. falciparum* infection which already cause significant mortality and morbidity.

Acknowledgements

I want to thank the Department of Hematology and Blood Transfusion at the University College Hospital, Ibadan.

Conflict of interest

The author declares no conflict of interest.

Author details

Chima Akunwata
University College Hospital, Ibadan, Nigeria

*Address all correspondence to: cubeautifulgates@gmail.com

References

[1] Reid ME, Bird GWG. Associations between human red cell blood group antigens and disease. Transfusion Medicine Reviews. 1990;**4**:47-55

[2] Abegaz SB, Human ABO. Blood groups and their associations with different diseases. BioMed Research International. 2021;**2021**:6629060. 9 pages. Available from: https://doi.org/10.115

[3] Höher G, Fiegenbaum M, Almeida S. Molecular basis of the Duffy blood group system. Blood Transfusion. 2018;**16**:93-100

[4] Driss A, Hibbert JM, Wilson NO, et al. Genetic polymorphisms linked to susceptibility to malaria. Malaria Journal. 2011;**10**:1-10

[5] Nkumama IN, Meara WPO, Osier FHA. Changes in malaria epidemiology in Africa and new challenges for elimination. Trends in Parasitology. 2017;**33**:128-140

[6] Cutbush M, Mollison P. The Duffy blood group system. Heredity (Edinb). 1950;**4**:383-389

[7] Cutbush M, Mollison P, Parkin D. A new human blood group. Nature. 1950;**165**:188-189

[8] Meny GM. The Duffy blood group system: A review. Immunohematology. 2010;**26**:51-56

[9] Marsh WL, Schmidt P. Present status of the Duffy blood group system. Critical Reviews in Clinical Laboratory Sciences. 1975;**5**:387-412

[10] Chaudhuri A, Polyakovat J, Zbrzeznat V, et al. Cloning of glycoprotein D cDNA, which encodes the major subunit of the Duffy blood group system and the receptor for the *Plasmodium vivax* malaria parasite. Proceedings of the National Academy of Sciences of the United States of America. 1993;**90**:10793-10797

[11] Pogo AO, Chaudhuri A. The Duffy protein: A malarial and chemokine receptor. Seminars in Hematology. 2000;**37**:122-129

[12] Langhi DM, Bordin JO. Duffy blood group and malaria. Hematology. 2006;**11**:389-398

[13] Darbonne WC, Rice GC, Mohler MA, et al. Red blood cells are a sink for interleukin 8, a leukocyte chemotaxin. The Journal of Clinical Investigation. 1991;**88**:1362-1369

[14] Castilho L. Duffy blood groups. In: Maloy S, Hughes K, editors. Brenner's Encyclopedia of Genetics. 2nd ed. London: Academic Press; 2013. pp. 425-427

[15] World Malaria Report 2020: 20 Years of Global Progress and Challenges. Geneva: World Health Organization; 2020

[16] Talapko J, Škrlec I, Alebi'c T, et al. Malaria: The past and the present. Microorganisms. 2019;**7**:179. DOI: 10.3390/microorganisms7060179

[17] Phillips MA, Burrows JN, Manyando C, et al. Malaria. Nature Reviews Disease Primers. 2017;**3**:1-24

[18] Khanam S. Prevalence and epidemiology of malaria in Nigeria: A review. International Research Journal of Pharmaceutical and Biosciences. 2017;**4**:10-12

[19] Price RN, Commons RJ, Battle KE, et al. *Plasmodium vivax* in the era of the shrinking *P. falciparum* map. Trends in Parasitology. 2020;**36**:560-570

[20] Fitri LE, Sardjono TW, Hermansyah B, et al. Unusual presentation of vivax malaria with anaemia, thrombocytopenia, jaundice,

renal disturbance, and melena: A report from Malang, a nonendemic area in Indonesia. Case Reports in Infectious Diseases. 2013;**2013**:686348. 4 pages. Available from: http://dx.doi.org/10.11

[21] Brazeau NF, Mitchell CL, Morgan AP, et al. The epidemiology of *Plasmodium vivax* among adults in the Democratic Republic of the Congo. Nature Communications. 2021;**12**:4169. DOI: 10.1038/s41467-021-24216-3

[22] Loy DE, Liu W, Li Y, et al. Out of Africa: Origins and evolution of the human malaria parasites *Plasmodium falciparum* and *Plasmodium vivax*. International Journal for Parasitology. 2018;**47**:87-97

[23] Battle EK, Baird JK. The global burden of *Plasmodium vivax* malaria is obscure and insidious. PLoS Medicine. 2021;**18**:e1003799

[24] Twohig KA, Pfeffer DA, Baird JK, et al. Growing evidence of *Plasmodium vivax* across malaria-endemic Africa. PLoS Neglected Tropical Diseases. 2019;**13**:1-16

[25] Poirier P, Lang CD, Atchade PS, et al. The hide and seek of *Plasmodium vivax* in West Africa: Report from a large-scale study in Beninese asymptomatic subjects. Malaria Journal. 2016;**15**:1-9

[26] Kwiatkowski DP. How malaria has affected the human genome and what human genetics can teach us about malaria. American Journal of Human Genetics. 2005;**77**:171-192

[27] Kano FS, de Souza AM, de Menezes TL, et al. Susceptibility to *Plasmodium vivax* malaria associated with DARC (Duffy antigen) polymorphisms is influenced by the time of exposure to malaria. Scientific Reports. 2018;**8**:13851. DOI: 10.1038/s41598-018-32254-z

[28] Tournamille C, Colin Y, Cartron J, et al. Disruption of a GATA motif in the Duffy gene promoter abolishes erythroid gene expression in Duffy-negative individuals. Nature Genetics. 1995;**70**:224-228

[29] Gao J, Chen YH, Peterson LC. GATA family transcriptional factors: Emerging suspects in hematologic disorders. Experimental Hematology & Oncology. 2015;**4**:28. DOI: 10.1186/s40164-015-0024-z

[30] McDermott DH. Cytokine receptor heterogeneity. In: Lotze MT, Thomso AW, editors. Measuring Immunity. London: Academic Press; 2005. pp. 23-34

[31] Zimmerman PA, Ferreira MU, Howes RE, et al. Red blood cell polymorphism and susceptibility to *Plasmodium vivax*. Advances in Parasitology. 2013;**81**:27-76

[32] Barnwell JW, Galinski MR. The adhesion of malaria merozoite proteins to erythrocytes: A reflection of function? Research in Immunology. 1991;**142**:666-672

[33] Galinski MŔ, Barnwell JW. *Plasmodium vivax*: Merozoites, invasion of reticulocytes and considerations for malaria vaccine development. Parasitology Today. 1996;**12**:20-29

[34] Aikawa M, Miller LH, Johnson J. Erythrocyte entry by malarial parasites a moving junction between erythrocyte and parasite. The Journal of Cell Biology. 1978;**77**:72-82

[35] Golassa L, Etego LA, Lo E, et al. The biology of unconventional invasion of Duffy-negative reticulocytes by *Plasmodium vivax* and its implication in malaria epidemiology and public health. Malaria Journal. 2020;**19**:1-10

[36] Aldarweesh FA. The Duffy blood group system. In: Erhabor O, Munshi A, editors. Human Blood Group System and Haemoglobinopathies. 2019. DOI: 10.5772/intechopen.89952

Chapter 5

Blood Groups: More than Inheritance of Antigenic Substances - Susceptibility to Some Diseases

Williams Bitty Azachi and Kuschak Mathias Dakop

Abstract

Blood group antigens represent polymorphic traits inherited among individuals and populations. The objective of this chapter is to review articles that have reported; the association between blood group antigens and susceptibility to some diseases. Findings showed that O blood group had a greater frequency of severe infections such as E coli, cholera and blood group A was associated with incidence of smallpox and some bacterial infections. These are principally based on presence or absence of "H-like" and "A and B-like" antigens markers. Antigens A, B and H are connected to N-glycans of vWF and reduces the half-life of the protein (10 hours) for group O while non-O groups, 25 hours. The loss of A, B, and H antigens as malignancy progresses was linked to potential metastasis. Similarly, some tumors have A or A-like antigens this explains the propensity of group A to develop tumors. Blood type incompatibility between mother and foetus sensitizes the mother to develop alloantibodies that could potentially cause death of the foetus in utero, a condition known hydrops. Reviewed articles have reported close link between blood group antigens and susceptibility diseases. More studies are required to rationalize the mechanism associated to this.

Keywords: blood group antigen, susceptible, disease

1. Introduction

Human blood groups since discovery in 1901 by Landsteiner have been widely studied and characterized. A total of 34 blood group systems have been recognized and documented by the International Society for Blood Transfusion (ISBT) [1, 2]. Each system is a series of red cell antigens, determined either by a single genetic locus or very closely linked loci. Alternative forms of a gene coding for red cell antigens at a particular locus are called alleles and individuals may inherited identical or non-identical alleles [3].

Many blood groups are receptors for toxins, parasites, and bacteria, where they can facilitate colonization or invasion or evade host clearance mechanisms [4]. Associations between the blood type and disease have been studied since the early 1900s when researchers determined that antibodies and antigens are inherited. However, due to lack of antigens of some blood groups, there have been some contentious issues with the association between the ABO blood group and vulnerability to certain infectious and non-infectious diseases [5].

ISBT no	Blood group system name	Blood group symbol	Chromosome
001	ABO	ABO	9
002	MNS	MNS	4
003	P	P1	22
004	Rh	RH	1
005	Lutheran	LU	19
006	Kell	KEL	7
007	Lewis	LE	19
008	Duffy	FY	1
009	Kidd	JK	18
010	Diego	DI	17
011	Yt or Cartwright	YT	7
012	XG	XG	X
013	Scianna	SC	1
014	Dombrock	DO	12
015	Colton	CO	7
016	Landsteiner-Weiner	LW	19
017	Chido/Rogers	CH/RG	6
018	Hh/Bombay	H	199
019	Kx	XK	X
020	Gerbich	GE	2
021	Cromer	CROM	1
022	Knops	KN	1
023	Indian	IN	11
024	Ok	OK	19
025	Raph	MER2	11
026	JMH	JMH	6
027	Li	I	6
028	Globoside	GLOB	3
029	GIL	GIL	9
030	Rh-associated glycoprotein	RHAG	6

IBST: International Society of Blood Transfusion, NO: Number.

Table 1.
Blood group systems recognized by the International Society of Blood Transfusion.

Fung et al. [1] (**Table 1**) gives blood group antigens characterized by ISBT.

2. ABO antigens linked to some diseases

2.1 Infectious disease

ABO gene products have been associated with some diseases [6–8]. The human body defense integrity against certain infections is characteristic of the presence or absence of blood group antigens and their corresponding antibodies. More so, data have shown that peptic ulceration was the first proven association to blood group gene products [9, 10]. *H pylori* is now known to be a causative agent leading to peptic ulceration and gastric cancer. *H pylori* has established colonies in the stomach of approximately one-half the world's population [11].

Red blood cell surface markers act as receptors for attachment to infectious agents and result in vulnerability difference among individuals with diverse receptor profile [12]. Some pathogens share genetic properties with their host. The relationship between ABO antigens and infections as vibrio cholera was discovered by early studies [12]. Major variations in ABO groups in the world were due to H-like antigen on the bacterium (*Vibrio cholera*) and an A-like antigen on Small pox virus [13]. This confers resistance status to people who make corresponding antibodies to H and A antigens [8]. Once a person gets infected with Cholera (Vibrio cholera strain O 1, E 1, Tor and O 139), the O blood types have a greater frequency of severe infections than the non-O blood types. Increased incidence of cholera is strongly associated to blood group O whereas, blood group A is strongly associated to smallpox, Pseudomoniasis, gonorrhea, tuberculosis and *streptococcus pneumoniae*, *Escherichia coli* and Salmonellosis. Blood type AB is linked to increased incidence of small pox, E. coli and Salmonellosis. Principally, these are associated to the presence of H antigen on group O and anti-H on group A and B [8].

The GI expresses Lewis and ABH antigens which is strongly linked to vulnerability of norovirus infection. People known as non-secretors are susceptible to infections caused by *Haemophilus influenzae*, Neisseria meningitides, *Streptococcus pneumoniae* and UTI caused by *E. coli*. ABO blood type is connected to peptic ulcer. The blood type O is highly susceptible to peptic ulcer than other blood type. *Helicobacter pylori* has been implicated to peptic ulcer similarly, *H. pylori* attachment to the human gastric mucosa was mediated by the H type 1 and Le b fucosylated antigens. Soluble glycoproteins of Le b inhibit *H. pylori* binding of *H. pylori*. This justifies the decreased infectivity commonly observed in the blood type A, B, and AB compared to blood group O [14–16].

2.2 Coronavirus

The novel virus, COVID 19 caused by SARS-CoV-2 widely spread around the globe is yet to be fully understood. Factors that influence susceptibility to the disease are age, sex, comorbid chronic disease etc. ABO blood group may influence the susceptibility to COVID. Blood group A have been linked with significant increase risk compared to blood group O due to like virus surface proteins. Blood group O persons can easily recognize these proteins as foreign and by extension confers lower chances to establish the disease [17]. Furthermore, anti-A inhibit binds of glycosylated SARS-CoV S protein expressing cells to angiotensin -converting enzyme 2 on cell membrane thereby truncate the interaction between the virus and its receptors, providing protection. Angiotensin converting enzyme activity is much in blood group B. This explains the possibility non-O blood group have more mortality [18, 19].

2.3 Coagulation

ABO blood types have been significantly linked to susceptibility to arterial and venous thromboembolism. There is an association between ABO antigens and the structural protein backbone of coagulation factors vWF and factor VIII which affects coagulation. Hypercoagulable plasma potentially causes venous thromboembolism and is characteristically observed in non- group O individuals due to higher levels of vWF. Von Willebrand factor is a large glycoprotein synthesized by Weibel-palade components in the endothelial cells and alpha granules of platelets. It is the carrier of factor VIII and plays a crucial role in plaletet adhesion and aggregation. Blood group O people have lower levels of vWF due to lack of additional carbohydrate to the terminal sugar. Moreso, plasma vWF is proteolyzed by metalloprotease enzyme ADAMTS13. This is faster in group O than non-O groups vVWF, thereby degrading FVIII levels. Group A, B and AB then have more vWF therefore increases FVIII levels [20].

2.4 Cardiovascular diseases

Blood group antigens do not cause cardiovascular diseases yet strongly linked to influence susceptibility. The known primary causes of cardiovascular diseases are genetic traits and life style among others. The ATP-binding cassette 2 genes are located at locus 9q34 which plays a significant role in cholesterol regulation. The H antigen has a connection to the structural back bone of coagulation factors (vWF and VIII) glycoproteins. This phenomenon explains greater risk non-O groups have for ischemic heart disease [8, 21].

Preeclampsia is serious condition with leading cause of intrauterine growth restrictions, maternal and foetal morbidity/mortality. Placental protein 13 is galectin that binds to beta-galactoside (N-acetyl-galactosamine, galactose, and fucose) linked to ABO antigens. This protein is observed in a pregnant woman with preeclampsia at early onset [22].

2.5 Malignancy

There are numerous publications in literature that have reported strong association between red cells antigens and some malignancies. ABH antigens are found on epithelial cells of the GIT, prostate, lungs, breast, uterine cervix, mouth, and bladder and their expression diminishes as malignancy progresses. Blood type antigens play a crucial role in cell signaling, cell recognition and cell adhesion yet these antigens are missing from the red cell membrane glycoprotein/glycolipids of malignant cells. This has been linked to DNA methylation in the promoter region. Blood group A gene in this case may inhibit transcription of the transferase enzyme with resultant loss of A antigen [15, 23]. Factually, malignancy progression results to loss of ABH antigens and potential metastasis of tumor cells. This phenomenon complicates routine red cell typing. Some tumor cells have been observed to mimic blood group A antigen markers giving group A person higher risk of disease progression than non-A blood groups [15]. It is important to note that blood group antigens do not cause malignancy rather susceptibility.

3. Association of other blood types and diseases

3.1 Haemolytic disease of the foetus and new-born (HDFN)

Haemolytic disease of the foetus and new born is a major clinical disease associated to Rhesus, Kell, Kidd and Duffy incompatibility between maternal alloantibody (IgG) to foetal antigens with resultant hemolysis of foetal red cells or suppression of the foetal red cell progenitors (commo with Kell system) [24]. Anti-D is the most implicated and severe form of HDFN, yet routine antenatal anti-D prophylaxis has ameliorated it.

3.2 Malaria

In western regions of Africa, it has been reported that individuals with negative Duffy blood type are common and confers protection against malaria caused by *Plasmodium vivax*. The resetting phenomenon commonly seen in parasitized erythrocytes. Forming rosettes is significantly lower in O blood types than blood group A red cells. It was observed that blood type A and B antigens are receptors for resetting on uninfected erythrocytes [25]. Parasitized RBC express resetting which helps us appreciate malaria pathogenesis. In blood group O, plasmodium falciparum invades Rbcs and little rosette is formed yet unstable which binds to uninfected Rbcs to form clusters of cells with resultant narrowing of vascular system. Some gene products such as pfEMP-1 and RIFIN secreted by Rbcs and parasite respectively have been implicated in ABO blood types and susceptibility to malaria among non-group O individuals. This could probably explain the ABO type commonly seen where malaria is prevalent.

4. Conclusion

It is fair to state that blood group antigens are not the primary cause of diseases but are associated to susceptibility to some diseases. Blood group antigens play a role as receptors or ligands to some disease processes. In general, non-O blood types are more susceptible to diseases than O.

Author details

Williams Bitty Azachi[1]* and Kuschak Mathias Dakop[2]

1 Tru-Care Medical Centre, Jos, Nigeria

2 Department of Medical Laboratory Science, University of Jos, Jos, Nigeria

*Address all correspondence to: wbitty4@gmail.com

References

[1] Fung MK, Grossman BJ, Hillyer CD, Westhoff CM. Technical manual. 18th ed. Bethesda, MD: AABB Press; 2014

[2] Reid ME, Lomas-Francis C, Olsson ML. Blood Group Antigen Facts Book. 3rd ed. Waltham, MA: Academic Press; 2012

[3] Lewis SM, Bain BJ, Bates I, Dacie JV. Dacie and Lewis Practical Haematology. Philadelphia: Churchill Livingstone/ Elsevier; 2012

[4] Cooling L. Blood groups in infection and host susceptibility. Clinical Microbiology Review. 2015;**28**(3): 801-870

[5] Abegaz SB. Human ABO blood groups and their associations with different diseases. BioMed Research International. 2021;**21**:1-9

[6] Boren T, Falk P, Roth KA, Larson G, Normark S. Attachment of helicobacter pylori to human gastric epithelium mediated by blood group antigens. Science. 1993;**262**(5141):1892-1895

[7] Cserti CM, Dzik WH. The ABO blood group system and plasmodium falciparum malaria. Blood. 2007;**110**(7):2250-2258

[8] Anstee DJ. The relationship between blood groups and disease. Blood. 2010;**115**(23):4635-4643

[9] Aird I, Bentall HH, Mehigan JA, Roberts JAF. The blood groups in relation to peptic ulceration and carcinoma of the colon, rectum, breast and bronchus. BMJ. 1954;**2**(4883):315

[10] Clarke CA, Cowan WK, Edwards JW, et al. The relationship of ABO blood groups to duodenal and gastric ulceration. BMJ. 1955;**2**(4940):643

[11] Bjorkholm B, Lundin A, Sillen A, et al. Comparison of genetic divergence and fitness between two subclones of helicobacter pylori. Infection and Immunity. 2001;**69**(12):7832-7838

[12] Cartron JP, Colin Y. Structural and functional diversity of blood group antigens. Transfusion Clinique et Biologique. 2001;**8**(3):163-199

[13] Garratty G. Relationship of Blood Groups to Disease: Do Blood Group Antigens Have a Biological Role? Southern, California: American Red Cross Blood Services; 2005

[14] Ewald R, Sumner S. Blood type biochemistry and human disease, "Wiley interdisciplinary reviews". Systems Biology and Medicine. 2016;**8**(6):517-535

[15] Yamamoto F, Cid E, Yamamoto M, Blancher A. ABO research in the modern era of genomics. Transfusion Medicine Reviews. 2012;**26**(2):103-118

[16] Garratty G. Blood groups and disease: A historical perspective. Transfusion Medicine Reviews. 2000;**14**(4):291-301

[17] Hari. Blood type and COVID-19 risk: O may help, a may hurt. The New England Journal of Medicine. 2020;**11**:325-329

[18] Laguipo ABB. Blood types and COVID-19 risk confirmed. The New England Journal of Medicine. 2020;**136**(18):1888

[19] Chung CM, Wang RY, Chen JW, et al. A genome-wide association study identifies new loci for ACE activity: Potential implications for response to ACE inhibitor. Pharmacogenomics Journal. 2020;**10**:537-544

[20] Franchini M, Capra F, Targher G, Montagnana M, Lippi, g. Relationship between ABO blood group and von Willebrand factor levels: From biology

to clinical implications. Thrombosis Journal. 2007;**5**:14

[21] Carpeggiani C, Coceani M, Landi P, Michelassi C, Abbate AL. ABO blood group alleles: A risk factor for coronary artery disease. An angiographic study. Atherosclerosis. 2010;**211**(2):461-466

[22] Than NG, Romero R, Meiri H, et al. P13, maternal ABO blood groups and the risk assessment of pregnancy complications. PLoS One. 2011;**6**(7): e21564

[23] Weisbrod AB, Nilubol N, Weinstein LS, et al. Association of Type-O Blood with neuroendocrine tumors in multiple endocrine neoplasia type 1. The Journal of Clinical Endocrinology & Metabolism. 2013;**98**(1):E109-E114

[24] Vaughan JI, Warwick R, Letsky E. Erythropoietic suppression in foetal anaemia because of Kell alloimmunization. American Journal of Obstetrics and Gynaecology. 1994;**171**(1):247-252

[25] Barragan A, Kremsner PG, Wahlgren M, et al. Blood group a antigen is a coreceptor in *plasmodium falciparum* resetting. Infectious Immunology. 2000;**68**:2971-2975

Chapter 6

RH Groups

Amr J. Halawani

Abstract

In 1939, a mother gave birth to a stillborn baby and underwent blood transfusion with ABO-matched blood from her husband. This resulted in a hemolytic transfusion reaction (HTR). Levine and Stetson postulated that a novel antigen was present in the baby and father, which was absent in the mother. Therefore, the mother's immune system recognized this antigen and produced antibodies against it. This condition has been known as the hemolytic disease of the newborn for a long period of time. Since the antenatal management of the fetus has been developed, the term has been modified to hemolytic disease of the fetus and newborn (HDFN). This case led to the discovery of the antibody against the first antigen of the RH blood group system, the D antigen. To date, 56 antigens have been recognized within the RH blood group system. The five main antigens are D, C, c, E, and e. As observed in the above-mentioned case, the antibodies against these antigens are implicated in HTR and HDFN.

Keywords: RH, anti-D, hemolytic disease of the fetus and newborn, antenatal and postnatal management, anti-D prophylaxis

1. Introduction

The RH blood group system is the most clinically significant blood group system after the ABO blood group system. It is extremely polymorphic, and to date, 56 antigens have been identified and reported by the International Society of Blood Transfusion [1]. The five main antigens of the RH blood group system are D, C, E, c, and e. Other antigens are represented in a combined form, including the ce or f antigen. Some antigens are correlated to specific ethnicities; e.g., the VS antigen is found in the Black population, which is a variant of the e antigen [2].

1.1 RH polypeptides

The antigens of the RH blood group system are encoded by two highly homologous genes, *RHD* and *RHCE*. The cDNA open reading frame of these 2 genes encodes 417 amino acids for each of the RhD and RhCE polypeptides, with a shared sequence identity of 92% (**Figure 1**). The difference between RhD and RhCE polypeptides is 32–35 amino acids, depending on which *RHCE* allele is inherited [*RHce*, *cE*, *Ce*, and *CE*] [3]. The RhD and RhCE polypeptides traverse the membrane lipid bilayer 12 times and form 6 extracellular loops, in which both NH_2 and COOH termini are intracellular [4]. In addition, the RhD and RhCE proteins may act as a possible CO_2 channel [5].

Normally, the *RHD* gene encodes for the D antigen, whereas the *RHCE* gene encodes for the C, c, E, and e antigens. Eight possible haplotypes have been identified, which vary from one population to another [6]. **Table 1** displays these

IntechOpen

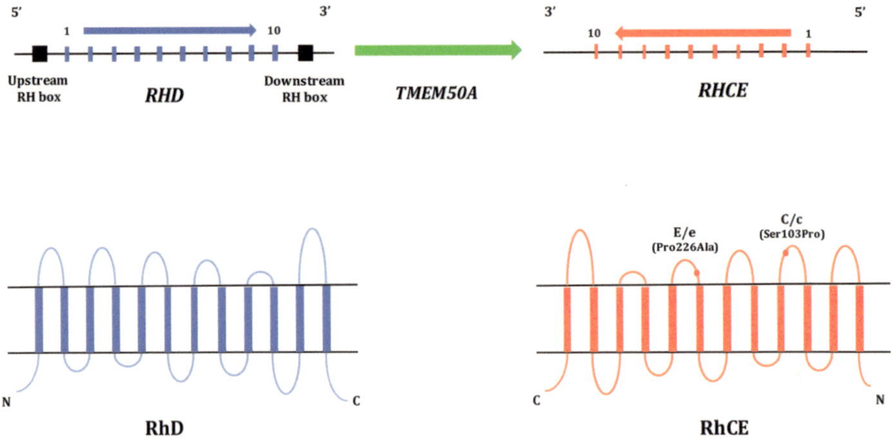

Figure 1.
The schematic diagram for the RH genes and their proteins. The RHD gene comprises 10 exons (blue boxes). Two RH boxes, upstream and downstream, are flanked by the RHD gene. The RHCE gene has the same number of exons (red boxes), but it is in the opposite orientation. The transmembrane 50A (TMEM50A) gene is flanked by the RHD and RHCE genes, which are indicated by a green arrow. Each protein, RhD and RhCE, traverses the red blood cell membrane 12 times forming 6 extracellular loops. On the RhCE protein, two amino acid substitutions, Ser103Pro and Pro226Ala, are indicated in the second and fourth extracellular loops of the RhCE protein. These gave rise to the antigenic polymorphism of the C/c and E/e antigens.

Haplotype	English [7]	Nigerian [8]	Chinese [9]	Saudi Arabian [10]
	N = 2000	N = 274	N = 4648	N = 3563
Dce (R_0)	0.0257	0.5908	0.0334	0.0078
DCe (R_1)	0.4205	0.0602	0.7298	0.4723
DcE (R_2)	0.1411	0.1151	0.1870	0.2736
DCE (R_Z)	0.0024	0	0.0041	0.0051
dce (r)	0.3886	0.2028	0.0232	0.2410
dCe (r')	0.0098	0.0311	0.0189	0
dcE (r'')	0.0119	0	0	0.0001
dCE (r^y)	0	0	0.0036	0

Table 1.
Frequencies of RH haplotypes in various ethnicities.

haplotypes along with the frequencies observed in different ethnicities. For instance, *Dce* is the most common haplotype observed in individuals of African origin compared with that observed in individuals from England and southwestern Saudi Arabia [7, 8, 10]. By contrast, the Chinese population lacks the *D* haplotypes [9].

The RhD/RhCE polypeptide is accompanied by two Rh-associated glycoproteins (RhAG) proteins. This association produces a trimer structure as a part of a macrocomplex on the red blood cell membrane proteins [11]. RhAG resembles the RhD and RhCE polypeptides with 36% identity and possesses glycosylation on the first loop. By contrast, the RhD and RhCE polypeptides are not glycosylated [12, 13].

1.2 RHD polymorphisms

The D+ antigen and D− are always denoted as Rh+ and Rh−, respectively. The presence of the D antigen in an individual means that their blood group is D+.

However, a person who lacks this antigen is considered D– [14]. Normal D+ individuals have a contact *RHD* gene with 10 contact exons without any mutations or modifications, with RH boxes that flank the *RHD* gene from upstream and downstream [15] (**Figure 1**). Regarding D– individuals, there are various genetic mechanisms underlying this phenotype according to ethnicity. For example, in Caucasians, the entire *RHD* gene in the *dce* haplotype is deleted, resulting in a hybrid box of both upstream and downstream RH boxes [16]. By contrast, Africans possess the pseudogene *RHDΨ*, which has a 37-bp duplication in intron 3 and exon 4, three missense mutations in exon 5, and a nonsense mutation with a premature stop codon in exon 6 [17].

Regarding the Asian population, D– is rare. Nevertheless, different mechanisms have been identified for the D– phenotype in this ethnicity, including the entire deletion of the RHD gene, D-elute (DEL) phenotype, and hybrid genes (such as *RHD-CE(2-9)-D* and *RHD-CE(3-9)-D*) [18–21].

1.3 Variants of the D antigen

1.3.1 Weak D

The weak D phenotype was previously designated as D^u because it could only be identified by anti-D immunoglobulin (Ig) G in the antiglobulin test and not with anti-D IgM. In contrast to the normal D antigen, the numbers of antigen sites per red blood cell are less and considered quantitative D [22–24]. The intact D antigen (normal D+) has 13,000–24,000 antigen sites per red blood cell. However, the weak D antigen possesses only between 70 and 4000 sites [25].

The weak D antigen comprises all the D epitopes but with weak expression. This phenotype arose from a missense mutation in the intracellular or intramembranous domain of the RhD polypeptides. Thus, a restriction occurs during the RhD polypeptide subunit assembly, leading to a decrease in the density of the RhD polypeptides [26]. In general, individuals with weak D cannot produce anti-D compared with partial D and are treated as D+ individuals [27]. However, in rare scenarios, weak D can produce anti-D. Hence, the term has been modified to weak partial D for such phenotypes. In summary, the term "D variants" was proposed by Daniels to be used for both weak D and partial D to clear the ambiguity [28]. A website called "The Human Rhesus Base" lists all D variant alleles [29, 30].

1.3.2 Partial D

The partial D phenotype was initially classified into six categories (i.e., I–VI) according to the patterns of antibody reactions with D+ red blood cells, which already produced anti-D [31, 32]. The development of monoclonal antibodies paved the way to identify different reaction patterns. The D antigen is now defined as a mosaic or made of pieces of "epitopes." Thirty epitopes have been identified and numbered as ep1–ep9, excluding ep7, followed by the subdivisions of these epitopes (e.g., ep8.3) [33].

This phenotype is characterized by the absence of some epitopes and is considered qualitative D [33]. Such individuals can produce anti-D when undergoing blood transfusion with a "complete" and intact normal D antigen and must be treated as D– phenotype when receiving a blood transfusion. In addition, a D– woman who is pregnant with a complete D+ child who inherits the paternal allele from his father is also at risk of developing hemolytic disease of the fetus and newborn (HDFN) [34]. The gene conversion of the two RH genes leads to the formation of a hybrid gene and the replacement of the *RHD* parts by the corresponding

RHCE ones. Furthermore, the presence of a missense mutation in the extracellular domains of the RhD polypeptides could result in the partial D phenotype [35].

1.4 Clinical significance of the RH groups

The most immunogenic antibody of the RH blood group system is the anti-D, which has been reported to cause severe hemolytic transfusion reaction (HTR). Therefore, typing for the D antigen is extremely crucial, except in a population in which D− is considered rare [36]. Of note, some D− individuals have been reported to produce anti-D antibodies when blood transfusion reaction occurs and D+ blood is transfused [37–39].

Anti-D has also been implicated in severe HDFN, leading to fetal mortality [40]. However, since the start of the use of anti-D prophylaxis, this issue has been decreased dramatically [41]. Other antibodies of the RH blood group system, namely, anti-C, anti-c, anti-E, and anti-e, have been observed to result in severe HTR and HDFN [42]. To date, more than 50 antigens among the RH and different blood group systems have been implicated in HDFN, ranging from mild to severe [43].

2. HDFN

2.1 HDFN pathophysiology

Maternal alloimmunization may be caused by blood transfusion with an incompatible blood group antigen or during the previous or present pregnancy, in which the fetus or neonate inherits the paternal allele of the blood group antigen that is different from the maternal allele [44]. Fetal red cell leakage via the placenta entering the maternal circulation is known as fetomaternal hemorrhage (FMH). These fetal red cells are recognized by the mother's immune system as foreign bodies and start producing IgM antibodies against these antigens. In the subsequent pregnancy, the maternal IgG antibodies cross the placenta, attach to the fetal red cells, and sensitize them. Consequently, alloimmune destruction can be triggered by splenic macrophages, resulting in anemia with erythroblastosis. The duration of hemolysis varies and causes antenatal or postnatal complications according to the development of the blood group antigen of the fetus [45]. Extramedullary erythropoiesis subsequently occurs to compensate for the destroyed red cells [46].

In the case of mild anemia, an appropriate compensation can be achieved by the liver and spleen. However, in complicated cases, severe anemia leads to hypoxia in multiple organs as a result of difficult delivery that requires sufficient oxygen and nutrients. This subsequently leads to circulatory and liver failure. Liver failure results in a decrease in protein levels and a drop in oncotic pressure in the circulation. Moreover, heart failure increases venous pressure. These two complications lead to ascites and edema, which is recognized as hydrops fetalis that has a high rate of mortality and stillbirth babies [47].

The level of HDFN severity varies from one fetus/infant to another [48]. In most severe cases, the fetus may die in the uterus starting from approximately week 17 of pregnancy [49]. In severe cases presenting with hydrops fetalis, which has high morbidity and mortality, patients could be diagnosed prior to or after the delivery. For the earlier detection of the disease, intrauterine transfusion can be performed in the patients [50]. For moderate HDFN, hyperbilirubinemia can be observed in patients. In such cases, exchange transfusion is needed for the neonate to prevent the accumulation of excess bilirubin in the brain, which may lead to neurological damage and kernicterus. Regarding mild cases, jaundice and hemolysis can be identified in the neonatal period [51].

The hemolysis of the red cells leads to the release of bilirubin into the fetal/neonatal circulation. This released bilirubin has the ability to traverse the placenta, which is then eliminated by the maternal liver. After delivery, the unconjugated bilirubin is processed and excreted by the neonate's liver. The enzyme diphosphate glucuronosyltransferase is released in a lesser amount owing to the immaturity of the neonate's liver [52].

Because of fetal hemoglobin destruction, all newborns can exhibit jaundice. Therefore, newborns with severe hemolysis have an increased level of bilirubin. This excess amount can accumulate in the brain and lead to irreversible central neurological damage and death, a condition known as kernicterus [53].

Anti-D is the major cause of HDFN, and it was a major cause of fetal and neonatal morbidity and mortality before 1970. The incidence of stillbirths and infant deaths dropped after anti-D prophylaxis was developed [54]. This anti-D prophylaxis inhibits the production of the maternal anti-D antibodies after the D-mismatch pregnancies. However, partial D mothers can develop anti-D antibodies when they become pregnant with D+ antigen fetus. In this regard, such mothers need to be treated as D− and should receive anti-D prophylaxis [55].

2.2 Factors affecting HDFN immunization and severity

Many factors affect the severity of RH-HDFN, including antigenic exposure, host factors, Ig class, antibody specificity, and influence of the ABO blood group.

2.2.1 Antigenic exposure

FMH from previous pregnancy can lead to a critical increase in maternal antibody titers, leading to maternal alloimmunization. FMH could occur in approximately 93% of mothers in a small amount (as low as 0.5 mL) [56]. The risk of FMH may be elevated as a result of abdominal trauma or because of interventions, including amniocentesis. Furthermore, the occurrence of FMH may exceed 50% at delivery owing to the entry of fetal red cells into the maternal circulation via placental separation from the uterus [57].

2.2.2 Host factors

With respect to the undefined genetic factors that lead to complications, the capability of antibody production differs in response to antigenic exposure [58]. For example, approximately 85% of D− individuals produce alloanti-D antibodies when transfused with blood containing D+ red cells. However, if anti-D prophylaxis has been administered to D− mothers, the remaining 16% will be at the risk of HDFN after pregnancy with D+ babies [59].

2.2.3 Ig class

Among the Ig classes (IgG, IgM, IgA, IgE, and IgD), only IgG has the ability to cross the placenta. The IgG1 and IgG3 subclasses are more efficient in causing red cell intravascular hemolysis compared with the IgG2 and IgG4 subclasses [60].

2.2.4 Antibody specificity

As previously described in this chapter, among all red cell antibodies, anti-D is the most immunogenic. Moreover, anti-C, anti-c, and anti-E are also potent antibodies that can cause moderate to severe HDFN. Anti-c and anti-E may lead to severe HDFN, which require management and treatment [61].

2.2.5 Influence of the ABO group

Of note, the incidence of detecting FMH is decreased when the mother is ABO-incompatible with the fetus. The rate of D immunization has been reported to be less in mothers with major ABO incompatibility with the fetus. This may be caused by the clearance of hemolysis that occurs due to the presence of ABO-incompatible D+ fetal red cells in the maternal circulation prior to the identification of fetal D+ red cells by the mother's immune system [57].

2.3 Anti-D prophylaxis

Anti-D prophylaxis is also known as RH immune globulin. In 1965, the success of the anti-D prophylaxis was reported; the prophylaxis was given to D– mothers after the delivery of the D+ newborn [62]. Anti-D prophylaxis has become the backbone therapy to preclude any clinical significance of RH-HDFN. However, the mode of action regarding how these prophylaxis works remain unclear [63].

Anti-D prophylaxis contains a natural product derived from human plasma that helps in the prevention of sensitization events; therefore, there might be a risk of infectious disease transmission [64]. In general, male blood donors are used and re-exposed to a small volume of the D antigen. As a consequence, the anti-D antibody is then developed by the immune system of the donors, followed by plasma collection and processing [65].

Anti-D prophylaxis was only administered to D– mothers postnatally, and in the 1970s, the prophylaxis was modified to include an antenatal dose for the additional prevention of any sensitizing event of RH-HDFN [66].

Different anti-D prophylaxis routines are used in different countries, in which most of them administer the antenatal dose at 34 weeks. The current regimen is by administering two doses to the D– or partial D mothers who are pregnant with D+ babies. The first dose of the anti-D prophylaxis should be administered to D– mothers within 28–34 weeks of pregnancy, which could decrease the antenatal immunization [67, 68]. A total of 92% of mothers were shown to have become sensitized after week 28 [64]. If sensitization events occur, anti-D prophylaxis must be immediately administered within 72 h [69].

The second dose of anti-D prophylaxis must be given within 72 h of delivery to all D– mothers. This dose has reduced the burden of RH-HDFN by approximately 95% in the last 52 years [70]. Indeed, the second dose was initially introduced in the first routine of anti-D prophylaxis [66]. The recommended dosage of anti-D is 300 mg [49]. A larger dose might be given depending on the magnitude of FMH to reduce the risk of sensitization. The dose is typically calculated depending on the magnitude of FMH either using the Kleihauer-Betke test or flow cytometry. Cellular assays might be performed to predict the severity of HDFN, including antibody-dependent cellular cytotoxicity assay, monocyte monolayer assay, and chemiluminescence [71].

2.4 Antenatal screening (fetus at risk)

In general, the recommended routine testing for ABO and D grouping on a maternal blood sample is performed in the first trimester to predict the severity of HDFN. Antibody screening is also performed. Antibody screening must be performed frequently during pregnancy to identify any emerging antibodies.

The early investigation of any new antibodies could assist in the monitoring and management of HDFN [72].

Maternal antibody titer testing is performed if IgG antibody screening is positive. Each antibody has a certain titer, which varies as per the antibody. If this titer is below a certain level, further management for HDFN is not necessary [73]. Titer measurement can also be beneficial to distinguish between immune and passive anti-D. Nevertheless, the severity of HDFN cannot be reliably determined [74]. If the maternal antibody level is higher than the critical titer, paternal *RHD* zygosity testing may be required.

2.4.1 Predicting D phenotype from DNA

2.4.1.1 Paternal RHD zygosity testing

The management of all immunized mothers is possible via paternal *RHD* zygosity testing, which detects the copy number of the father's gene. This identifies the father's zygosity status and paternity. Nowadays, various techniques have been used for *RHD* zygosity testing. These include real-time polymerase chain reaction (PCR), digital PCR, and mass spectrometry [75–78].

When fathers are homozygous for the deletion of the entire *RHD* gene, i.e., D−, there is no risk of HDFN within the current or subsequent pregnancies. By contrast, when fathers are D+ carrying both the alleles of positive *RHD* gene (homozygous, e.g., *DCe*/*DCe*), then the fetus will definitely be D+. The fetus has a 50% possibility of being D+ in case the father's genotype is hemizygous (e.g., *DCe*/*dce*). Therefore, further analysis using noninvasive prenatal testing (NIPT) is required [77].

Some methods assess the hybrid Rhesus box expressed in D− individuals, which results from the entire *RHD* gene deletion in Caucasians. It mainly targets the *cde* haplotype; therefore, such a method cannot be applied to the samples of African ethnicity [79]. A total of 59% of Nigerians [8] possess the *Dce* haplotype [see **Table 1**].

2.4.1.2 Fetal genotyping

Amniocentesis is a risky and invasive procedure for obtaining fetal DNA. It can increase the risk of alloimmunization and miscarriage. Instead, NIPT is nowadays performed for fetal genotyping. The approach is now followed globally and is performed by detecting cell-free fetal DNA extracted from the maternal plasma [80, 81].

Fetal genotyping is a noninvasive technique based on cell-free fetal DNA circulating in the maternal plasma, which is derived from a maternal blood sample. This approach assists in identifying different blood group alleles and predicting the corresponding antigens, including D, C, c, E, and e, in addition to KEL1 antigens [82, 83].

If the mother is D− and the fetus is D+, anti-D prophylaxis may be administered. By contrast, if the fetus is D−,the D− mother does not need to receive the prophylaxis. Therefore, this avoids unnecessary injections in the mother; moreover, it appears to be cost-effective to save treatment for the mothers in actual need [84]. Thus, the approach of fetal genotyping could be cost-effective [85].

2.5 Antenatal management

2.5.1 Ultrasonography

Amniocentesis has been replaced by the middle cerebral artery (MCA) Doppler ultrasonography for predicting the severity of fetal anemia. At present, this method is routinely used by obstetricians for investigating fetal anemia by observing an increase in the velocity of blood flow in MCA in anemic fetuses compared with normal ones [86, 87]. In the case of severe anemia, fetal blood sampling (FBS) can be performed along with cordocentesis. Under ultrasound guidance, FBS is normally performed with a needle to obtain fetal blood; this procedure is considered an invasive procedure [88].

2.5.2 Intrauterine transfusion

For severely HDFN-affected fetuses, intrauterine transfusion provides blood to the fetus via the umbilical vein under ultrasound guidance. Fetuses must be administered O– blood (unless fetal ABO type is known), which is also KEL-1 negative, leuko depleted, plasma depleted, hemoglobin S negative, cytomegalovirus seronegative, and ≤5 days old [89]. Furthermore, citrate phosphate dextrose is the anticoagulant used in these blood units to prevent problems arising from using different anticoagulants. Gamma irradiation is performed to eliminate any residual leukocytes that may lead to graft-versus-host disease [89].

Fetal hemoglobin and hematocrit level measurement along with crossmatching is performed to ensure that safe compatible blood is transfused to the fetus. All blood units are normally prewarmed to 37°C before being transfused. Interestingly, intrauterine transfusion has been reported to reduce prenatal death and stillbirth resulting from HDFN by 75–90% [90, 91].

2.6 Postnatal screening (newborn/infant at risk)

The serological investigation is performed after a sample is withdrawn from the cord blood at birth. This sample is used to detect HDFN, which may help arrange for probable transfusion.

2.6.1 ABO grouping

The forward ABO blood group typing may be observed with weak reactions with anti-A and anti-B antisera in infants compared with adults and older children [57]. This is because the ABO antigens of newborn infants are not entirely developed and may take 5–10 years to reach the adult levels [92]. Furthermore, reverse ABO grouping is not feasible because infants do not produce ABO antibodies at that age.

2.6.2 D typing

In rare cases, an infant's red cells strongly bind to the maternal anti-D antibody, resulting in a false-negative type of D or what is known as blocked RH [93]. Anti-D can be identified from the eluant of these red cells, and typing these eluted red blood cells should be observed for any reaction with the anti-D antibody.

2.6.3 Direct antiglobulin test (DAT)

DAT is very crucial in diagnosing HDFN. A positive DAT demonstrates a sensitization reaction in which an infant's red cells are coated by maternal IgG antibodies. No correlation has been observed between the severity of HDFN and the reaction strength. Moreover, other laboratory or clinical manifestations for hemolysis can also lead to positive DAT. This could be owing to the mother receiving the anti-D prophylaxis [57].

2.6.4 Elution

Performing the elution test in the case of positive DAT is not essential as a routine procedure. As previously mentioned, eluant is the solution for blocked RH [57, 93].

2.7 Infant management

2.7.1 Phototherapy

This procedure is used to treat anemic newborns with mild to moderate HDFN and who have elevated levels of bilirubin. Phototherapy is performed using a blue-green light, with wavelength ranging from 460 to 490 nm [94]. Natural direct sunlight has been reported to have the same wavelength and can be beneficial to reduce jaundice, although it is not recommended because it may increase the risk of sunburn [95, 96].

Bilirubin is a lipophilic molecule that absorbs light and is metabolized into two isomers, which are less lipophilic (in other words, water-soluble) and less toxic to the brain. These can then be excreted via the urine without the requirement of enzymatic glucuronidation [97]. Overall, phototherapy is an effective procedure and can adequately conjugate bilirubin. Furthermore, it may assist reduce the requirement of blood transfusion [57].

2.7.2 Exchange transfusion

Exchange transfusion may be performed in newborns who demonstrate severe anemia with hyperbilirubinemia or heart failure. During pregnancy, the fetal liver is unable to metabolize the unconjugated bilirubin. Therefore, this unconjugated bilirubin, made by the fetus, is crossed the placenta and metabolized by the maternal disposal system. After delivery, this system is no longer used as well as the infant's liver is immature and cannot metabolize the unconjugated bilirubin efficiently. Therefore, the high level of bilirubin in the newborn may lead to kernicterus, which is the accumulation of bilirubin in the brain.

The assessment of hemoglobin and bilirubin levels is essential to determine the requirement of exchange transfusion in neonates. This is performed for removing bilirubin and maternal antibodies from neonates [98, 99]. Exchange transfusion is indicated at a critical level of bilirubin (i.e., ≥100 μmol/L), depending on the neonatal age.

Exchange transfusion is the replacement of neonatal blood by whole blood or equivalent, with the concurrent removal of bilirubin and maternal antibodies. However, this procedure is labor-intensive and time-consuming. Therefore, its use has become rare owing to the use of anti-D prophylaxis and phototherapy [50].

2.7.3 Red cell transfusion

Infants may receive red cell transfusion immediately after delivery for several weeks to treat severe anemia. The same criteria that are used for intrauterine transfusion and exchange transfusion must be applied for blood transfusion. Newborns must be closely monitored for any clinical signs of ongoing anemia, particularly if the infant is malnourished or sleeps heavily [98].

3. Conclusion

In summary, RH is the most highly variable blood group system. The antigens of this system have many variants that include the D variants such as weak D and partial D. The antibodies of this system can cause HTR due to incompatibility during a blood transfusion. Furthermore, their antibodies can lead to fetal red cell sensitization and destruction, causing moderate to severe HDFN. Nowadays, owing to the development of the latest technologies, the antenatal management of fetuses has become feasible along with postnatal management. Anti-D prophylaxis can now be administered antenatally at gestational weeks 28–34. At-risk pregnancies can also be monitored noninvasively using MCA ultrasonography and fetal genotyping.

Conflict of interest

The author declares no conflict of interest.

Author details

Amr J. Halawani
Faculty of Applied Medical Sciences, Department of Laboratory Medicine, Umm Al-Qura University, Makkah, Saudi Arabia

*Address all correspondence to: ajjhalawani@uqu.edu.sa

References

[1] International Society of Blood Transfusion. Red cell immunogenetics and blood group terminology [Internet]. 2021. Available from: https://www.isbtweb.org/working-parties/red-cell-immunogenetics-and-blood-group-terminology [Accessed: September 06, 2021]

[2] Chou ST, Westhoff CM. The Rh and RhAG blood group systems. Immunohematology. 2010;**26**:178-186. Available from: https://pubmed.ncbi.nlm.nih.gov/22356455/

[3] Mouro I, Colin Y, Chérif-Zahar B, Cartron JP, Le Van Kim C. Molecular genetic basis of the human rhesus blood group system. Nature Genetics. 1993;**5**:62-65. DOI: 10.1038/ng0993-62

[4] Chérif-Zahar B, Bloy C, Le Van Kim C, Blanchard D, Bailly P, Hermand P, et al. Molecular cloning and protein structure of a human blood group Rh polypeptide. Proceedings of the National Academy of Sciences of the United States of America. 1990;**87**:6243-6247. DOI: 10.1073/pnas.87.16.6243

[5] Soupene E, King N, Feild E, Liu P, Niyogi KK, Huang CH, et al. Rhesus expression in a green alga is regulated by CO_2. Proceedings of the National Academy of Sciences of the United States of America. 2002;**99**:7769-7773. DOI: 10.1073/pnas.112225599

[6] Daniels G. Rh and RHAG blood group systems. In: Human Blood Groups. 3rd ed. UK: Wiley-Blackwell; 2013. DOI: 10.1002/9781118493595.ch5

[7] Race RR, Mourant AE, Lawler SD, Sanger R. The Rh chromosome frequencies in England. Blood. 1948;**3**:689-695. DOI: 10.1182/blood.V3.6.689.689

[8] Mourant AE, Domaniewska-Sobczak K, Kopec AC. The Distribution of the Human Blood Groups and Other Polymorphisms. 2nd ed. London: Oxford University Press; 1976. Available from: http://lib.ugent.be/catalog/rug01:000041013

[9] Mackay J, Wang JC, Wong KS. The incidence of blood groups in 4648 southern Chinese from Toishan district and its vicinity. Journal of Hong Kong Medical Technology Association. 1969;**1**:11-14

[10] Halawani AJ, Arjan AH. ABO, RH, and KEL1 antigens, phenotypes and haplotypes in southwestern Saudi Arabia. Clinical Laboratory. 2021;**67**: 344-348. DOI: 10.7754/Clin.Lab.2020.200633

[11] Conroy MJ, Bullough PA, Merrick M, Avent ND. Modelling the human rhesus proteins: Implications for structure and function. British Journal of Haematology. 2005;**131**:543-551. DOI: 10.1111/j.1365-2141.2005.05786.x

[12] Eyers SA, Ridgwell K, Mawby WJ, Tanner MJ. Topology and organization of human Rh (rhesus) blood group-related polypeptides. The Journal of Biological Chemistry. 1994;**269**:6417-6423. DOI: 10.1016/S0021-9258(17)37388-X

[13] Marini AM, Urrestarazu A, Beauwens R, André B. The Rh (rhesus) blood group polypeptides are related to NH_4^+ transporters. Trends in Biochemical Sciences. 1997;**22**:460-461. DOI: 10.1016/s0968-0004(97)01132-8

[14] Colin Y, Chérif-Zahar B, Le Van Kim C, Raynal V, Van Huffel V, Cartron JP. Genetic basis of the RhD-positive and RhD-negative blood group polymorphism as determined by Southern analysis. Blood. 1991;**78**:2747-2752. DOI: 10.1182/blood.V78.10.2747.2747

[15] Chérif-Zahar B, Le Van Kim C, Rouillac C, Raynal V, Cartron JP, Colin Y. Organization of the gene (RHCE) encoding the human blood group RhCcEe antigens and characterization of the promoter region. Genomics. 1994;**19**:68-74. DOI: 10.1006/geno.1994.1014

[16] Wagner FF, Flegel WA. RHD gene deletion occurred in the rhesus box. Blood. 2000;**95**:3662-3668. DOI: 10.1182/blood.V95.12.3662

[17] Singleton BK, Green CA, Avent ND, Martin PG, Smart E, Daka A, et al. The presence of an RHD pseudogene containing a 37 base pair duplication and a nonsense mutation in Africans with the Rh D-negative blood group phenotype. Blood. 2000;**95**:12-18. DOI: 10.1182/blood.V95.1.12

[18] Shao CP, Maas JH, Su YQ, Köhler M, Legler TJ. Molecular background of Rh D-positive, D-negative, D(el) and weak D phenotypes in Chinese. Vox Sanguinis. 2002;**83**:156-161. DOI: 10.1046/j.1423-0410.2002.00192.x

[19] Sun CF, Chou CS, Lai NC, Wang WT. RHD gene polymorphisms among RhD-negative Chinese in Taiwan. Vox Sanguinis. 1998;**75**:52-57. DOI: 10.1159/000030958

[20] Peng CT, Shih MC, Liu TC, Lin IL, Jaung SJ, Chang JG. Molecular basis for the RhD negative phenotype in Chinese. International Journal of Molecular Medicine. 2003;**11**:515-521. DOI: 10.3892/ijmm.11.4.515

[21] Xu Q, Grootkerk-Tax MG, Maaskant-van Wijk PA, Van Der Schoot CE. Systemic analysis and zygosity determination of the RHD gene in a D-negative Chinese Han population reveals a novel D-negative RHD gene. Vox Sanguinis. 2005;**88**:35-40. DOI: 10.1111/j.1423-0410.2005.00584.x

[22] Bush M, Sabo B, Stroup M, Masouredis SP. Red cell D antigen sites and titration scores in a family with weak and normal Du phenotypes inherited from a homozygous Du mother. Transfusion. 1974;**14**:433-439. DOI: 10.1111/j.1537-2995.1974.tb04557.x

[23] Cunningham NA, Zola AP, Hui HL, Taylor LM, Green FA. Binding characteristics of anti-Rh0 (D) antibodies to Rh0 (D)-positive and Du red cells. Blood. 1985;**66**:765-768. DOI: 10.1182/blood.V66.4.765.765

[24] Szymanski IO, Araszkiewicz P. Quantitative studies on the D antigen of red cells with the Du phenotype. Transfusion. 1989;**29**:103-105. DOI: 10.1046/j.1537-2995.1989.29289146825.x

[25] Wagner FF, Frohmajer A, Ladewig B, Eicher NI, Lonicer CB, Müller TH, et al. Weak D alleles express distinct phenotypes. Blood. 2000;**95**:2699-2708. DOI: 10.1182/blood.V95.8.2699

[26] Wagner FF, Gassner C, Müller TH, Schönitzer D, Schunter F, Flegel WA. Molecular basis of weak D phenotypes. Presented at the 25th Congress of the International Society of Blood Transfusion held in Oslo on June 29, 1998 and published in abstract form in Vox Sang 74:55, 1998[suppl]. Blood. 1999;**93**(1):385-393. DOI: 10.1182/blood.V93.1.385

[27] Sandler SG, Chen LN, Flegel WA. Serological weak D phenotypes: A review and guidance for interpreting the RhD blood type using the RHD genotype. British Journal of Haematology. 2017;**179**:10-19. DOI: 10.1111/bjh.14757

[28] Daniels G. Variants of RhD--current testing and clinical consequences. British Journal of Haematology. 2013;**161**:461-470. DOI: 10.1111/bjh.12275

[29] The human RhesusBase, version 2.5. Update. 2020. Available from: http://www.rhesusbase.info/

[30] Wagner FF, Flegel WA. The rhesus site. Transfusion Medicine and Hemotherapy. 2014;**41**:357-363. DOI: 10.1159/000366176

[31] Tippett P, Sanger R. Observations on subdivisions of the Rh antigen D. Vox Sanguinis. 1962;7:9-13. DOI: 10.1111/j.1423-0410.1962.tb03223.x

[32] Tippett P. Further observations on subdivisions of the Rh antigen D. 1977. Available from: http://pascal-francis.inist.fr/vibad/index.php?action=getRecordDetail&idt=PASCAL7850148289

[33] Scott ML, Voak D, Jones JW, Avent ND, Liu W, Hughes-Jones N, et al. A structural model for 30 Rh D epitopes based on serological and DNA sequence data from partial D phenotypes. Transfusion Clinique et Biologique. 1996;**3**:391-396. DOI: 10.1016/S1246-7820(96)80051-6

[34] Argall CI, Ball JM, Trentelman E. Presence of anti-D antibody in the serum of a Du patient. The Journal of Laboratory and Clinical Medicine. 1953;**41**:895-898. DOI: 10.5555/uri:pii:0022214353900730

[35] Flegel WA. Molecular genetics of RH and its clinical application. Transfusion Clinique et Biologique. 2006;**13**:4-12. DOI: 10.1016/j.tracli.2006.02.011

[36] Lin M. Taiwan experience suggests that RhD typing for blood transfusion is unnecessary in Southeast Asian populations. Transfusion. 2006;**46**:95-98. DOI: 10.1111/j.1537-2995.2006.00680.x

[37] Frohn C, Dümbgen L, Brand JM, Görg S, Luhm J, Kirchner H. Probability of anti-D development in D− patients receiving D+ RBCs. Transfusion. 2003;**43**:893-898. DOI: 10.1046/j.1537-2995.2003.00394.x

[38] Yazer MH, Triulzi DJ. Detection of anti-D in D−recipients transfused with D+ red blood cells. Transfusion. 2007;**47**:2197-2201. DOI: 10.1111/j.1537-2995.2007.01446.x

[39] Gonzalez-Porras JR, Graciani IF, Perez-Simon JA, Martin-Sanchez J, Encinas C, Conde MP, et al. Prospective evaluation of a transfusion policy of D+ red blood cells into D− patients. Transfusion. 2008;**48**:1318-1324. DOI: 10.1111/j.1537-2995.2008.01700.x

[40] Levine P, Stetson RE. An unusual case of intra-group agglutination. Journal of the American Medical Association. 1939;**113**:126-127. DOI: 10.1001/jama.1939.72800270002007a

[41] Freda VJ, Gorman JG, Pollack W. Rh factor: Prevention of isoimmunization and clinical trial on mothers. Science. 1966;**151**:828-830. DOI: 10.1126/science.151.3712.828

[42] Koelewijn JM, Vrijkotte TG, Van Der Schoot CE, Bonsel GJ, De Haas M. Effect of screening for red cell antibodies, other than anti-D, to detect hemolytic disease of the fetus and newborn: A population study in the Netherlands. Transfusion. 2008;**48**:941-952. DOI: 10.1111/j.1537-2995.2007.01625.x

[43] Moise KJ. Hemolytic disease of the fetus and newborn. Clinical Advances in Hematology & Oncology. 2013; **11**:664-666

[44] De Haas M, Thurik FF, Koelewijn JM, Van der Schoot CE. Haemolytic disease of the fetus and newborn. Vox Sanguinis. 2015;**109**:99-113. DOI: 10.1111/vox.12265

[45] Lopriore E, Rath ME, Liley H, Smits-Wintjens VE. Improving the management and outcome in haemolytic disease of the foetus and newborn. Blood Transfusion. 2013;**11**:484

[46] Urbaniak SJ, Greiss MA. RhD haemolytic disease of the fetus and the

newborn. Blood Reviews. 2000;**14**:44-61. DOI: 10.1054/blre.1999.0123

[47] Smits-Wintjens VE, Walther FJ, Lopriore E. Rhesus haemolytic disease of the newborn: Postnatal management, associated morbidity and long-term outcome. Seminars in Fetal and Neonatal Medicine. 2008;**13**:265-271. DOI: 10.1016/j.siny.2008.02.005

[48] Dean L. Chapter 4, Hemolytic disease of the newborn. In: Blood Groups and Red Cell Antigens [Internet]. Bethesda, US: National Center for Biotechnology Information; 2005. Available from: https://www.ncbi.nlm.nih.gov/books/NBK2266/

[49] Klein HG, Anstee DJ. Chapter 12—Haemolytic disease of the foetus and newborn. In: Mollison's Blood Transfusion in Clinical Medicine. 12th ed. UK: John Wiley & Sons, Wiley-Blackwell; 2014

[50] Jackson ME, Baker JM. Hemolytic disease of the fetus and newborn: Historical and current state. Clinics in Laboratory Medicine. 2021;**41**:133-151. DOI: 10.1016/j.cll.2020.10.009

[51] Akkök ÇA, Seghatchian J. Pediatric red cell and platelet transfusions. Transfusion and Apheresis Science. 2018;**57**:358-362. DOI: 10.1016/j.transci.2018.05.019

[52] Mitchell S, James A. Severe late anemia of hemolytic disease of the newborn. Paediatrics & Child Health. 1999;**4**:201-203. DOI: 10.1093/pch/4.3.201

[53] Leung AK, Sauve RS. Breastfeeding and breast milk jaundice. Journal of the Royal Society of Health. 1989;**109**:213-217. DOI: 10.1177/146642408910900615

[54] Tovey LA. Oliver memorial lecture. Towards the conquest of Rh haemolytic disease: Britain's contribution and the role of serendipity. Transfusion Medicine. 1992;**2**:99-109. DOI: 10.1111/j.1365-3148.1992.tb00142.x

[55] Lukacevic Krstic JL, Dajak S, Bingulac-Popovic J, Dogic V, Mratinovic-Mikulandra J. Anti-D antibodies in pregnant D variant antigen carriers initially typed as RhD+. Transfusion Medicine and Hemotherapy. 2016;**43**:419-424. DOI: 10.1159/000446816

[56] Sebring ES, Polesky HF. Fetomaternal hemorrhage: Incidence, risk factors, time of occurrence, and clinical effects. Transfusion. 1990;**30**:344-357. DOI: 10.1046/j.1537-2995.1990.30490273444.x

[57] Harmening DM. Hemolytic disease of the fetus and newborn (HDFN). In: Modern Blood Banking & Transfusion Practices. 7th ed; Philadelphia: FA. Davis Company; 2018

[58] Higgins JM, Sloan SR. Stochastic modeling of human RBC alloimmunization: Evidence for a distinct population of immunologic responders. Blood. 2008;**112**:2546-2553. DOI: 10.1182/blood-2008-03-146415

[59] Ayache S, Herman JH. Prevention of D sensitization after mismatched transfusion of blood components: Toward optimal use of RhIG. Transfusion. 2008;**48**:1990-1999. DOI: 10.1111/j.1537-2995.2008.01800.x

[60] Singh A, Solanki A, Chaudhary R. Demonstration of IgG subclass (IgG1 and IgG3) in patients with positive direct antiglobulin tests. Immunohematology. 2014;**30**:24-27

[61] Babinszki A, Berkowitz RL. Haemolytic disease of the newborn caused by anti-c, anti-E and anti-Fya antibodies: Report of five cases. Prenatal Diagnosis. 1999;**19**:533-536. DOI: 10.1002/(SICI)1097-0223(199906)19:6<533::AID-PD570>3.0.CO;2-5

[62] Freda VJ, Gorman JG, Pollack W. Successful prevention of experimental Rh sensitization in man with an anti-Rh gamma 2-globulin antibody preparation: A preliminary report. Transfusion. 1964;**4**:26-32. DOI: 10.1111/j.1537-2995.1964.tb02824.x

[63] Kumpel BM, Elson CJ. Mechanism of anti-D-mediated immune suppression–a paradox awaiting resolution? Trends in Immunology. 2001;**22**:26-31. DOI: 10.1016/S1471-4906(00)01801-9

[64] Bowman JM. RhD hemolytic disease of the newborn. The New England Journal of Medicine. 1998;**339**:1775-1777. DOI: 10.1056/NEJM199812103392410

[65] Kumpel BM. Monoclonal anti-D for prophylaxis of RhD haemolytic disease of the newborn. Transfusion Clinique et Biologique. 1997;**4**:351-356. DOI: 10.1016/s1246-7820(97)80040-7

[66] Liumbruno GM, D'Alessandro A, Rea F, Piccinini V, Catalano L, Calizzani G, et al. The role of antenatal immunoprophylaxis in the prevention of maternal-foetal anti-Rh (D) alloimmunisation. Blood Transfusion. 2010;**8**:8-16. DOI: 10.2450/2009.0108-09

[67] Jones ML, Wray J, Wight J, Chilcott J, Forman K, Tappenden P, et al. A review of the clinical effectiveness of routine antenatal anti-D prophylaxis for rhesus-negative women who are pregnant. BJOG : An International Journal of Obstetrics and Gynaecology. 2004;**111**:892-902. DOI: 10.1111/j.1471-0528.2004.00243.x

[68] Kumar S, Regan F. Management of pregnancies with RhD alloimmunisation. BMJ. 2005;**330**:1255-1258. DOI: 10.1136/bmj.330.7502.1255

[69] Crowther CA, Middleton P. Anti-D administration after childbirth for preventing rhesus alloimmunisation. Cochrane Database of Systematic Reviews. 2000;**2**:CD000021. DOI: 10.1002/14651858.CD000021

[70] Weatherall DJ. Cyril Clarke and the prevention of rhesus haemolytic disease of the newborn. British Journal of Haematology. 2012;**157**:41-46. DOI: 10.1111/j.1365-2141.2012.09031.x

[71] Hadley AG. Laboratory assays for predicting the severity of haemolytic disease of the fetus and newborn. Transplant Immunology. 2002;**10**:191-198. DOI: 10.1016/S0966-3274(02)00065-5

[72] Fung MK, Grossman BJ, Hillyer CD, Westhoff CM, editors. Technical Manual. 18th ed. Bethesda: American Association of Blood Banks Press; 2014

[73] Bennardello F, Coluzzi S, Curciarello G, Todros T, Villa S, Italian Society of Transfusion Medicine and Immunohaematology (SIMTI) and Italian Society of Gynaecology and Obstetrics (SIGO) Working Group. Recommendations for the prevention and treatment of haemolytic disease of the foetus and newborn. Blood Transfusion. 2015;**13**:109-134. DOI: 10.2450/2014.0119-14

[74] Gottvall T, Hllden JO. Concentration of anti-D antibodies in Rh (D) alloimmunized pregnant women, as a predictor of anemia and/or hyperbilirubinemia in their newborn infants. Acta Obstetricia et Gynecologica Scandinavica. 1997;**76**:733-738. DOI: 10.3109/00016349709024338

[75] Chiu RW, Murphy MF, Fidler C, Zee BC, Wainscoat JS, Lo YM. Determination of RhD zygosity: Comparison of a double amplification refractory mutation system approach and a multiplex real-time quantitative PCR approach. Clinical Chemistry. 2001;**47**:667-672. DOI: 10.1093/clinchem/47.4.667

[76] Krog GR, Clausen FB, Dziegiel MH. Quantitation of RHD by real-time polymerase chain reaction for determination of RHD zygosity and RHD mosaicism/chimerism: An evaluation of four quantitative methods. Transfusion. 2007;**47**:715-722. DOI: 10.1111/j.1537-2995.2007.01175.x

[77] Sillence KA, Halawani AJ, Tounsi WA, Clarke KA, Kiernan M, Madgett TE, et al. Rapid RHD zygosity determination using digital PCR. Clinical Chemistry. 2017;**63**:1388-1397. DOI: 10.1373/clinchem.2016.268698

[78] Gassner C, Meyer S, Frey BM, Vollmert C. Matrix-assisted laser desorption/ionisation, time-of-flight mass spectrometry–based blood group genotyping—the alternative approach. Transfusion Medicine Reviews. 2013;**27**:2-9. DOI: 10.1016/j.tmrv.2012.10.001

[79] Grootkerk-Tax MG, Maaskant-van Wijk PA, Van Drunen J, Van Der Schoot CE. The highly variable RH locus in nonwhite persons hampers RHD zygosity determination but yields more insight into RH-related evolutionary events. Transfusion. 2005;**45**:327-337. DOI: 10.1111/j.1537-2995.2005.04199.x

[80] Norwitz ER, Levy B. Noninvasive prenatal testing: The future is now. Reviews in Obstetrics and Gynecology. 2013;**6**:48-62

[81] Chandrasekharan S, Minear MA, Hung A, Allyse MA. Noninvasive prenatal testing goes global. Science Translational Medicine. 2014;**6**(6):231fs15. DOI: 10.1126/scitranslmed.3008704

[82] Finning K, Martin P, Summers J, Daniels G. Fetal genotyping for the K (Kell) and Rh C, c, and E blood groups on cell-free fetal DNA in maternal plasma. Transfusion. 2007;**47**:2126-2133. DOI: 10.1111/j.1537-2995.2007.01437.x

[83] Finning KM, Martin PG, Soothill PW, Avent ND. Prediction of fetal D status from maternal plasma: Introduction of a new noninvasive fetal RHD genotyping service. Transfusion. 2002;**42**:1079-1085. DOI: 10.1046/j.1537-2995.2002.00165.x

[84] Tiblad E, Taune Wikman A, Ajne G, Blanck A, Jansson Y, Karlsson A, et al. Targeted routine antenatal anti-D prophylaxis in the prevention of RhD immunisation-outcome of a new antenatal screening and prevention program. PLoS One. 2013;**8**:e70984. DOI: 10.1371/journal.pone.0070984

[85] Teitelbaum L, Metcalfe A, Clarke G, Parboosingh JS, Wilson RD, Johnson JM. Costs and benefits of non-invasive fetal RhD determination. Ultrasound in Obstetrics & Gynecology. 2015;**45**:84-88. DOI: 10.1002/uog.14723

[86] El Shourbagy S, Elsakhawy M. Prediction of fetal anemia by middle cerebral artery Doppler. Middle East Fertility Society Journal. 2012;**17**:275-282. DOI: 10.1016/j.mefs.2012.09.003

[87] Mari G, Deter RL, Carpenter RL, Rahman F, Zimmerman R, Moise KJ Jr, et al. Noninvasive diagnosis by Doppler ultrasonography of fetal anemia due to maternal red-cell alloimmunization. The New England Journal of Medicine. 2000;**342**:9-14. DOI: 10.1056/NEJM200001063420102

[88] Avent ND. Haemolytic disease of the fetus and newborn. In: Transfusion and Transplantation Science. UK: Oxford University Press; 2018. Available from: https://books.google.com.sa/books?id=A_okvgAACAAJ

[89] Oepkes D, van Scheltema PA. Intrauterine fetal transfusions in the management of fetal anemia and fetal thrombocytopenia. Seminars in Fetal & Neonatal Medicine. 2007;**12**:432-438. DOI: 10.1016/j.siny.2007.06.007

[90] Rodeck CH, Holman CA, Karnicki J, Kemp JR, Whitmore DN, Austin MA. Direct intravascular fetal blood transfusion by fetoscopy in severe rhesus isoimmunisation. Lancet. 1981;**317**:625-627. DOI: 10.1016/S0140-6736(81)91549-X

[91] Kanhai HH, Bennebroek Gravenhorst JB, Van Kamp IL, Meerman RH, Brand A, Dohmen-Feld MW, et al. Management of severe hemolytic disease with ultrasound-guided intravascular fetal transfusions. Vox Sanguinis. 1990;**59**:180-184. DOI: 10.1111/j.1423-0410.1990.tb00855.x

[92] Auf der Maur CA, Hodel M, Nydegger UE. Rieben RAge dependency of ABO histo-blood group antibodies: Reexamination of an old dogma. Transfusion. 1993;**33**:915-918. DOI: 10.1046/j.1537-2995.1993.331194082382.x

[93] Sulochana PV, Rajesh A, Mathai J, Sathyabhama S. Blocked D phenomenon, a rare condition with Rh D haemolytic disease of newborn–a case report. International Journal of Laboratory Hematology. 2008;**30**:244-247. DOI: 10.1111/j.1751-553X.2007.00943.x

[94] Ebbesen F, Hansen TWR, Maisels MJ. Update on phototherapy in jaundiced neonates. Current Pediatric Reviews. 2017;**13**:176-180. DOI: 10.2174/1573396313666170718150056

[95] Salih FM. Can sunlight replace phototherapy units in the treatment of neonatal jaundice? An in vitro study. Photodermatology, Photoimmunology & Photomedicine. 2001;**17**:272-277. DOI: 10.1034/j.1600-0781.2001.170605.x

[96] Slusher TM, Olusanya BO, Vreman HJ, Wong RJ, Brearley AM, Vaucher YE, et al. Treatment of neonatal jaundice with filtered sunlight in Nigerian neonates: Study protocol of a non-inferiority, randomized controlled trial. Trials. 2013;**14**:446. DOI: 10.1186/1745-6215-14-446

[97] Bhutani VK, Stark AR, Lazzeroni LC, Poland R, Gourley GR, Kazmierczak S, et al. Predischarge screening for severe neonatal hyperbilirubinemia identifies infants who need phototherapy. The Journal of Pediatrics. 2013;**162**:477-482.e1. DOI: 10.1016/j.jpeds.2012.08.022

[98] Rath ME, Smits-Wintjens VE, Walther FJ, Lopriore E. Hematological morbidity and management in neonates with hemolytic disease due to red cell alloimmunization. Early Human Development. 2011;**87**:583-588. DOI: 10.1016/j.earlhumdev.2011.07.010

[99] American Academy of Pediatrics Subcommittee on Hyperbilirubinemia. Management of hyperbilirubinemia in the newborn infant 35 or more weeks of gestation. Pediatrics. 2004;**114**:297-316. DOI: 10.1542/peds.114.1.297

Chapter 7

Neutrophil-Specific Antigens: Immunobiology, Genetics and Roles in Clinical Disorders

Parviz Lalezari and Behnaz Bayat

Abstract

Neutrophils are the most abundant nucleated cells in blood circulation and play important roles in the innate and adaptive immune responses. Neutrophil-specific antigens, only expressed on neutrophils, are glycoproteins originally identified in studies on neonatal neutropenia due to fetal-maternal incompatibility and autoimmune neutropenia of infancy. The most investigated neutrophil–specific antigens are the NA and NB antigens that their incompatibilities also cause transfusion-induced febrile reactions and acute lung injury, a potentially fatal reaction, and in bone marrow transplantation, causing graft rejection. NA antigens are members of the immunoglobulin superfamily and are low-affinity Fc-receptors FcγRIIIb (CD16b). Fc receptors connect the F(ab), the antigen-binding fragment of the antibody molecules, to neutrophils and lead them to recognize and phagocytize the targeted antigens. The NB (CD177) antigen belongs to the urokinase-type Plasminogen Activator Receptor Superfamily (uPAR, CD59, Ly6), but its specific functions have not been fully determined. It is known, however, that NB antigen binds proteinase-3 (PR3 to the neutrophil membrane), a serine protease. In clinical studies, it was also demonstrated that NB expression is highly elevated in Polycythemia Vera and is unexpectedly expressed in some cancer tissues. Neutrophil-specific antigens are examples of antigens that have important biological and clinical activities beyond antigenicity.

Keywords: alloimmune neonatal neutropenia (ANN), autoimmune neutropenia of infancy (AINI), neutrophil-specific antigens, NA, SH, NB1, NB2, Fc gamma Receptor IIIB (FcγRIIB), HLA antigens, human neutrophil antigen (HNA) nomenclature

1. Introduction

Neutrophils are the most abundant circulating leukocytes. They are short-lived, terminally differentiated myeloid cells produced in the bone marrow. They contain intracellular granules and vesicles and a multi-lobed nucleus. The condensed nuclear chromatin is considered to indicate a limited transcriptional activity in the cells. Neutrophils are an important part of innate and adaptive immunity, and the first line of defense against invading pathogens. There are multiple types of proteins expressed on neutrophil surfaces, including Fc-receptors, adhesion molecules, integrins, multiple cytokine receptors, innate immune receptors such as Toll-like receptors, and C-type lectins. Some receptors directly recognize and bind to pathogens and play their role in the defensive functions, some other receptors, such as Fc-receptors, bind the Fc fragment of the antibody molecules to neutrophils, which enable the neutrophils to recognize the targets of the antibodies and then perform their functions in adaptive immunity. Some other receptors recognize the inflammatory signals. The extracellular structures from neutrophil surface proteins are involved in signal transduction, cell-cell interactions, neutrophil migration, and substance releases such as cytokines, reactive oxygen species, exocytosis of intracellular granules, and neutrophil extracellular trap (NET) formation, the structure that traps and kills the invading bacteria. Some of the glycoproteins expressed on the neutrophil surface carry the Neutrophil-Specific Antigens. The protective effects of neutrophils are also contributed by their ability to receive signals through their receptors that stimulate them to rapidly generate pseudopods, move and invade their targets. They phagocytize the microorganisms and digest them by proteolytic enzymes present in their intracellular granules. Eli Metchnikoff (1845–1916) of Pasteur Institute in Paris was the first investigator who identified neutrophil immunogenicity. He reported that guinea pigs injected with rabbit blood produce antibodies against rabbit leukocytes and called the antibodies leukoagglutinins and leukolysins [1]. In 1926, Charles Doan recognized that leukocyte incompatibility causes transfusion reaction and suggested that leukocytes have their own "groups" [2]. Jean Dausset later reported that the presence of leukoagglutinins in blood could be associated with chronic autoimmune neutropenia [3]. In 1957, Tom Brittingham, of the hematology team of Washington University in St. Louis, produced leukoagglutinins experimentally [4]. He volunteered to receive weekly 100 mL blood for nine consecutive weeks from a patient who had chronic myelogenous leukemia, the last four causing chills and fever. He could prevent the reaction by minimizing the number of injected leukocytes. To study the effects of antibodies in donors' blood, he received 50 mL blood from a patient who had received numerous transfusions. This injection caused high fever, vomiting, dyspnea, cyanosis and hypotension. A chest X-ray on the following day revealed bilateral lung infiltrations, a reaction that is now recognized as transfusion-related acute lung injury (TRALI). Dausset [5] and others [6, 7] in subsequent years demonstrated inheritance of leukocyte antigens.

2. Alloimmune neonatal neutropenia (ANN)

Neutrophil-specific antigens were identified during the investigation on alloimmune neonatal neutropenia (ANN), proven to be due to fetal-maternal incompatibility [8]. In that study, the first child had died from neutropenia and sepsis, and the three subsequent infants also had neutropenia at birth. The diagnosis of ANN was based on the detection of a strong leukoagglutinin in the maternal serum that reacted with the paternal neutrophils but not with the maternal own cells. The blood smear from the last infant showed the absence of mature neutrophils but lymphocytes, monocytes, eosinophils, basophils and platelets were normal. Neutrophil-specificity of the antibody was established by demonstrating that the antibody could be absorbed only by isolated neutrophils and not by any other blood cells or tissue cells [9]. An anti-5b antibody [10], known to react with all blood leukocytes and many other tissue cells tested, was used as a control and showed that the ANN antibody reacted only with neutrophils. This first neutrophil-specific antigen identified was named NA1, "N" for Neutrophil, "A" for the first locus, and 1 for the first allele. Further investigation on ANN led to the identification of other antigens including NA2, NC1 [9, 11], SH [12, 13] and HNA-1d [14]. ANN could be identified because the tests were directly performed on the blood from the affected infants and their parents, and not on blood samples from multiparous female donors who develop multiple forms of antibodies against the HLA, red blood cells and platelets. This may explain the reason for some other investigators inability to recognize ANN, and even conclude that neonatal neutropenia was the result of infection and not alloimmunization [15]. Neutrophil specificity causes ANN because the total neutrophil mass in fetus blood is small, and there are no other cell types in the body that express the same antigens and can absorb the maternal antibodies. The more common HLA antibodies in maternal blood do not cause ANN because HLA antigens are widely distributed in blood cells and other tissues, and also, as soluble antigens, can absorb the maternal-derived antibodies [16]. In one ANN example, both neutrophil-specific and HLA antibodies were detected in the maternal plasma, but only the neutrophil-specific antibody was detected in the affected newborn's blood. ANN is a self-limited disorder and lasts as long as the maternal antibodies remain in the infant's circulation, a period of a few weeks to a few months. The recovery can be predicted when the antibody can no longer be detected in the infant's plasma. In ANN, the absolute neutrophil counts are below 500 and the common symptoms are omphalitis, and infections in the skin and in other locations. It has been estimated that ANN occurs in 0.1–0.2% of pregnancies. This may be an underestimation; because now most birth deliveries are performed under sterile conditions and the newborns are not exposed to bacteria, and consequently, neutropenic infants may remain asymptomatic and undiagnosed. Also, blood counts and differentials are not routinely tested on asymptomatic newborns. ANN, in contrast to erythroblastosis fetalis, known as Rh disease, can occur in the first pregnancies. This information should warn the blood banks that blood donors with single pregnancy and asymptomatic children or undiagnosed ANN infants may be carriers of the hazardous neutrophil-specific antibodies, and transfusion of their blood may result in severe reactions and even fatalities.

2.1 Autoimmune neutropenia of infancy (AINI)

The second clinical disorder that increased interest in neutrophil-specific antigens was autoimmune neutropenia of infancy (AINI). In 1975, a strong anti-NA2 antibody was identified in the blood of a six months old infant. The antibody was not present in the mother's blood and the child was too old to have ANN [17]. Additional studies [18] demonstrated autoantibodies on 119 of 121 severely neutropenic children. The disorder, called AINI, was shown to be very common and the autoantibodies were directed against neutrophil-specific antigens found in ANN. Thus far, the antigens NA, NB, ND1 [19] NE1 [20] and 9a [18] are associated with AINI, and some of these antibodies are also found in a few cases of autoimmune neutropenia in adults [16]. Before detection of autoantibodies and establishing the autoimmune nature of this disorder, AINI was recognized as "chronic benign neutropenia," benign because it was a self-limited neutropenia disorder, lasting several months to a few years. The neutrophil count in blood of AINI patients is between zero to 500, and there is an elevation of monocytes and eosinophils, which may represent a protective defense mechanism. Pathogenesis of AINI has not yet been determined, but it has been suggested that a delay in maturation of the regulatory T-cells (T-regs) causes a lack of tolerance to neutrophil-specific antigens, and that the autoimmunity becomes corrected when T-regs become functional [21, 22].

3. Neutrophil-specific antigens

The NA and NB antigen systems are the most investigated neutrophil-specific antigens and are discussed here, with their information summarized in

Tables 1 and **2**. The genetics and biological functions of these two systems are different and are described separately. However, their clinical impacts, such as their roles in ANN, AINI, blood transfusion, and bone marrow transplantation are similar. Also, in this report, the nomenclature used is based on the original and not the HNA that will be discussed separately.

3.1 The NA system

Four expressing alleles, NA1, NA2, NA3 (SH), and NA4 are currently known in this system (**Table 1**). NA1 was the first, described with neonatal neutropenia. The antigen SH, here referred to as NA3, was first described in one case of ANN, and subsequently in three other cases [12, 13]. The antigen referred to as NA4, was originally described in two cases of ANN [14]. In rare individuals, the NA antigens are not expressed. This abnormality, caused by gene alteration, is called NA^{null} [12, 23, 24]. The NA antigens appear at the metamyelocytic phase of neutrophil maturation and are anchored on the cell membranes by the Glycosylphosphatydylinositol (GPI) at the density of 100,000 to 300,000 copies per cell [25, 26]. Some NA molecules are also stored inside the cells and translocate to the membrane and are released during cell activation. NA antigens are the low-affinity Fc-receptors, FcγRIII [27]. Fc-receptors are glycoproteins that bind the Fc fragment of immunoglobulins and connect the humoral to the cellular components of the immune system. In this process, the effector cell, via its Fc-receptor, is connected to the F(ab)2, the antigen-binding part of the antibodies. This connection leads to phagocytosis and antibody-mediated cell cytotoxicity (ADC) [28]. The genetic polymorphisms in the coding genes for Fc-receptors influence their effectiveness. Three Fc gamma receptor (FcγR) subclasses have been identified thus far: FcγRI, FcγRII, and FcγRIII [29]. The NA epitopes are located on FcγRIIIb (CD16b), exclusively expressed on neutrophils, bind Fc fragments of IgG1 and IgG3 immunoglobulin subclasses [30, 31]. The *CD16B* gene has 5 exons and is located on chromosome 1q23.3, and it belongs to the immunoglobulin superfamily. FcγRIIIb contains two immunoglobulin G (IgG)-like domains and the proximal domain carries the IgG-binding segments. The four variations in *FCGR3B* coding region lead to the formation of the four known NA alleles (**Tables 1** and **2**). These SNPs are located on exon 3 of the *FCGR3B* gene. *FCGR3B*01* and *FCGR3B*02* are the most frequent alleles coding NA1 and NA2 respectively [32]. These two alleles are

System	Allele	Frequency	Location	Amino acids	MW (kDa)*	Chromosome	Exons	Nucleotide, base pairs
NA (CD16b)	NA1	0.560	Membrane, intracellular	233	58–65	1q23.3	5	699
	NA2	0.885		233	65–80			
	NA3 (SH)	0.05		233	65–80			
	NA4 (HNA1d)	Unknown		233	65–80			
	NA^{null}	Rare						
NB (CD177)	NB1	0.97	Membrane, secondary granules, vesicles	437	55–64	19q13.2	9; 6 in the pseudo-gene	1614

**Difference in molecular weight (MW) is due to differences in glycosylation.*

Table 1.
Neutrophil-specific antigens, the NA and NB system.

System	Gene	Protein	Antigen name	Nucleotide	Amino acids
FcγRIIIb (CD16)	*FCGR3B*	CD16	NA1	108G, 114C, 194A, 233C, 244G, 316G	36R, 38 L, 65 N, 78A, 82D, 106 V
			NA2	108C, 114 T, 194G, 233A/C, 244A, 316A	36S, 38 L, 65S, 78D/A, 82 N, 106I
			NA3 (SH)	108C, 114 T, 194G, 233A, 244A, 316A	36S, 38 L, 65S, 78D, 82 N, 106I
			NA4 (HNA1-d)	108C,114 T,194G, 233A, 244A, 316A	36S, 38 L, 655S, 78A, 82 N,106I
			NA^{null}	—	—
CD177	*CD177*	CD177	NB1	787A	Full length
CD177 negative	*CD177* altered	NB2(?)*	NB2(?)* NB1-neg	787 T	264 (Truncated length)
CD177	*CD177* modified	NB1 soluble in plasma	NB1	1291G > A	Truncated at the GPI-binding site

**NB2 is suggested to be a truncated protein, but needs to be confirmed.*

Table 2.
Neutrophil-specific antigens alleles; nucleotide and amino acid differences.

determined by five nucleotides in positions 108, 114, 194, 244, and 316, and except for a silent position in position 114, the other SNPs result in the replacement in positions 36, 65, 82, and 106 of IgG-like domains distal to the cell membrane [33]. The presence of asparagine in position 65 is determinant for the formation of the NA1 epitope. For the formation of the NA2 epitope, the presence of amino acid Ser in position 36 and Asn in position 82 is necessary. Therefore, to predict the NA phenotype, it is necessary to determine the FCGRIIIB genotype in positions 108 and 244. Since no evidence exists regarding the involvement of amino acid 106 in the NA epitope formation, it does not necessary to determine the FCGRIIIB genotype in position 316 [33]. A polymorphism in nucleotide c.233 of *FCGR3B*02* is responsible for the generation of the third allele, *FCGR3B*03* known as NA3 (SH). This SNP changes the amino acid alanine in position 78 to Asp [13]. The glycoprotein encoded by the *FCGRB*03* allele bears both epitopes for NA2 and NA3. The fourth *FCGR3B* allele, *FCGR3B*04* encoded by *FCGR3B*01* is formed by an SNP c.316G > A, changing amino acids in positions p.106Val > Ile. NA4 epitope is coded by *FCGR3B*02*. Alloantibodies against NA4 were produced by individuals typed *FCGR3B*01/FCGR3B*03* (NA1/NA3) against NA2 coded by *FCGR3B*02*. Therefore, immunization against the NA4 is only formed in NA1/NA3 individuals against NA2 individuals [14]. NA1 and NA2 are different in molecular weight caused by disparate N-glycosylation sites: 4 glycosylation sites for NA1 and 6 glycosylation sites for NA2 [34]. In some individuals in the European population (about 0.15% of the normal population), the *FCGR3B* gene, is missing with no detectable FcγRIIIb on the neutrophil surface [35]. Accordingly, some individuals carry three copies of *FCGR3B* genes [36]. FcγRIIIb contributes to phagocytosis, elimination of immune complexes, and antibody-mediated cell cytotoxicity [37]. The co-localization of FcγRIIIb with CD11b/CD18 on lipid rafts participates in signal transduction involved in neutrophil activation and production of reactive oxygen species (ROS) [31]. A clinical study on individuals with NA^{null} phenotypes has found only four out of 21 subjects to develop frequent infections [38]. Although, NA^{null} individuals are mostly healthy without symptoms, their exposure to NA positive cells during pregnancy, after transfusion or transplantation induces immunization and consequently the production of iso-antibodies against NA antigen (s). The clinical relevance of NA antigens other than ANN, AINI and autoimmunity include TRALI, febrile transfusion reactions and bone marrow transplantation incompatibility issues. NA2 in some populations has been documented as a susceptibility to the development of systemic lupus erythematosus (SLE) [39]. Also, NA antigens are known to be lost in Paroxysmal Nocturnal hemoglobinurea (PNH) [40], because of loss of GPI in PNH. A separate gene, *CD16A* discovered by Retch [32], produces another Fc-receptor that has medium affinity, and also binds IgG1 and IgG3 immunoglobulins. These two proteins (FcγRIIIa and FcγRIIIb) are highly homologous; however, *FCGR3A* codes a protein with 21 transmembrane extra amino acid and is longer than *FCGR3B* and is expressed on natural killer cells, a subset of monocytes, macrophages, and T-cells, but not on neutrophils [41]. The amino acid differences in the four NA expressing alleles are NA1 and NA2 differ in three amino acids, and NA3 and NA4 are variations of NA2. NC1, as previously described [11], has been reported to be related to, or identical to NA2 [42]. Neutrophil stimulation induces a proteolytic cleavage and releases membrane FcγRIIIb molecules to plasma [43].

3.2 NB antigen

NB1 [44–46], also known as CD177, is a cysteine-rich glycoprotein that expresses at the promyelocytic phase of neutrophil maturation [47]. This early expression on immature neutrophils causes the bone marrow of the newborns affected with NB1 alloimmune neonatal neutropenia, to show a "maturation arrest" profile, and even be misdiagnosed as acute leukemia. NB1 is anchored on the neutrophil membrane, and also mediates the expression of proteinase 3 (PR3), a serine protease enzyme, on the neutrophil membrane [48]. NB1 is also present in neutrophil secondary granules and intracellular vesicles. Stroncek and Skubitz determined that NB1 is a 58–65 KDa, GPI anchored glycoprotein [49, 50]. Not all circulating neutrophils in NB1-positive individuals carry the NB1 antigen on their surfaces [19, 51, 52]. This bimodal expression divides the neutrophils into NB1-positive and NB1-negative subpopulations in NB1-positive blood. Although, the percentage of NB1-positive neutrophils remains stable, infection, pregnancy, treatment with granulocyte-colony stimulating factor (G-CSF), and Polycythemia Rubra Vera (PRV) upregulate, CD177 expression [49, 53, 54]. CD177 is not detectable on neutrophils of 3–5% of normal populations. These individuals are defined as CD177-null. Kissel [55] sequenced the gene and determined it to be composed of 1614 base pair nucleotides, belonging to the urokinase-type Plasminogen Activator Receptor Superfamily (uPAR, CD59, Ly6), located on chromosome 19q13.3 [53]. The gene has 9 exons and a pseudogene composed of 6 exons derived from the original gene and is reversely positioned [52, 56–58]. A missense mutation in exon 7 of CD177 gene, c.787A > T replaces amino acid 263 with a stop codon and induces production of a truncated protein and consequently loss of CD177 expression on the neutrophil surface (**Table 2**) [57]. The stop codon responsible for the absence of CD177 protein on the neutrophil surface arises when exon 7 in the CD177 coding region is provided by the CD177 pseudogene, called CD177P1 [52]. The heritable ratio between CD177/CD177P1 determines CD177 high and low expression; individuals homozygous for the CD177 gene have higher CD177 expression whereas the existence of CD177P1 sequence in the CD177 gene leads to the presence of CD177 negative subpopulation [52]. However, in only 40% of CD177-null individuals, the presence of c.787 T homozygote is responsible for the absence of CD177 protein from the neutrophil surface [58, 59]. Later studies have added c.1291G > A SNP that affects the GPI anchor region of CD177 molecule and converts membrane-bound to the soluble form of CD177. This polymorphism has been introduced as a genetic regulator for atypical/low expression that participates in the mechanism of CD177 deficiency. The combination of SNPs, c.787 T and c.1291A is responsible to regulate the presence of CD177 on the neutrophil surface [60]. A study on the epigenetic component that regulates CD177 expression, has documented a non-classical random monoallelic expression (MAE) on neutrophil subsets of CD177 positive individuals [61]. The complete absence of gene transcription (neither complete, nor truncated) in the CD177 negative neutrophil subpopulation has been introduced as a mechanism regulating the absence of CD177 protein on the CD177 negative subpopulation. However, later study analyzing mRNA content in sorted CD177 positive and negative subpopulations has shown the presence of CD177 mRNA in both neutrophil subsets and doubted the previous observation [59]. Additional interpretation on the potential role of c.787 T and c.1291 on NB system is presented in the NB2–dilemma subsection, separately.

The role of CD177 on neutrophil function is complicated and poorly understood. Physical association of CD177 with CD11b/CD18, concentrated on lipid rafts, support transduction of signals initiated after binding of anti-PR3 antibodies (such as Anti-Neutrophilic Cytoplasmic Autoantibody (ANCA)) on CD177/PR3 complex toward neutrophil cytoplasm and consequently leads to neutrophil activation, degranulation, and superoxide production [62]. The association of CD177 in ANCA-mediated neutrophil degranulation may explain the selective activation of CD177 positive neutrophil subsets involved in the mechanism of ANCA-associated vasculitis (AAV). Analysis has introduced an epitope formed by CD177/PR3 complex as a relevant epitope for CD177 autoantibodies but not isoantibodies. Interestingly this epitope is a signaling relevant epitope that upon binding of relevant antibodies induces neutrophil activation [63]. A previous study indicated CD177 as a heterophilic binding partner for platelet endothelial cell adhesion molecule 1 (PECAM-1). Inhibitory analysis has introduced PECAM-1 membrane-proximal domain 6 to mediate the binding of CD177. The later study however introduced PR3 as a binding partner for PECAM-1 but not CD177 alone [63]. The CD177/PECAM-1 interaction is considered to induce a signal in endothelial cells, destabilize VE-cadherin from the endothelial junction, and lead to the preferential trans-endothelial migration ability of CD177-positive neutrophils [64, 65]. The binding of monoclonal antibodies to CD177 (such as MEM166) enhances the expression of β2 integrins, activates CD177 positive neutrophils, raises neutrophil adhesion, and interrupts neutrophil's migratory ability [66]. This effect rather than direct CD177/PECAM-1 binding, explains a neutrophil adhesion via a CD177-driven pathway [67]. In CD177 deficient mice, besides a slight decrease in neutrophil counts, no defect in neutrophil function, chemotaxis, and clearance of bacterial infection was observed [68]. Although, no distinct difference between CD177 positive and negative neutrophil subsets have been described and both CD177 positive and null individuals are healthy, in multiple diseases the CD177 expression/upregulation has been introduced as a risk factor. In comparison to healthy donors, a higher proportion of CD177 positive neutrophils have been detected in patients with ANCA vasculitis [69]. The study documented a differential gene expression between CD177 negative and positive neutrophils; CD177 positive produces a lower level of pro-inflammatory cytokines and exhibits increased bactericidal activities such as ROS production and neutrophil extracellular trap (NET) [70]. A decreased percentage of CD177 positive neutrophils in Myelodysplastic Syndrome has been documented [71]. In contrast, in Kawasaki disease, an epigenetic hypomethylation and consequently increased CD177 gene transcription has been reported [72]. Beyond neutrophils, the presence of CD177 on epithelial cells, in cervical cancer, prostate cancer and ovarian cancer has been documented. Analysis of CD177 in mammary epithelial cells revealed a strong CD177 expression on epithelial cells that was significantly reduced in paired cancer specimens [73]. These data suggest CD177 molecule as a cancer cell-intrinsic tumor suppressor and introduced this molecule as a potential good prognostic factor in different cancer types [73]. A recent study documented TRALI induction in a CD177 pre-immunized recipient after transfusion with packed red blood cells (PRBC) from CD177 positive donors, but not CD177 negative donors. In vitro analysis has documented the presence of soluble CD177/PR3 complex in plasma from CD177 donors that was significantly increased after PRBC filtration. The molecular mechanism regulating the secretion of CD177 in plasma is yet to be resolved. PECAM-1 on pre-activated endothelial cells absorbs soluble CD177/PR3 complex from plasma. Binding of CD177 isoantibodies to the absorbed CD177/PR3 complex on endothelial cells, induced endothelial barrier dysfunction implicated in the mechanism of anti-CD177 mediated reverse TRALI [74]. In severe cases of COVID-19 infections, progressive respiratory failure results from immunothrombosis that is driven by activated neutrophils and platelets. As an activation marker, CD177 molecules were upregulated in severe COVID-19 cases [75]. Further studies have documented a correlation between CD177 upregulation in the serum and disease severity and mortality in COVID-19 [76].

3.3 NB2 dilemma

The presence of a stop codon on the CD177 gene is shown to prevent NB1 biosynthesis [77]. Based on this observation, the NB-negative individuals are called NB^{null}. This interpretation also suggests that there is no allele to NB1. In contrast to this conclusion, a NB1 positive mother was reported to have a child with ANN and have a neutrophil-specific antibody that reacted only with NB1-negative neutrophils. This antibody could be absorbed only by NB1-negative neutrophils and was called anti-NB2, an allele to NB1 [78]. In this case, the maternal neutrophils were also found to be 9a-negative, which suggested a possible relationship between NB2 and 9a, an antigen previously described by van Rood. Also, examination of data from testing neutrophils from 76 members of 11 families, tested with anti-9a and anti-NB2 showed identical results in 72 of 76 donors (these results were obtained during 1967 HL-A workshop experiments in Torino, Italy). Anti-NB2 antibodies were also reported to cause febrile transfusion reaction [79] and TRALI [80]. Future studies are needed to confirm that NB2, an allele to NB1, exists or if its identification has been the result of serological errors. If confirmed, NB2 would be a truncated NB1 molecule in NB-negative individuals who carry the c.787 stop codon, which causes the absence of amino acids responsible for CD177 expression (**Table 2**). Therefore, NB2 will be CD177-negative but not NB^{null}. It should also be recognized that individuals with SNP c.1291 G > A expression have soluble NB1 antigens present in their plasma and not on their neutrophils, and thus will be classified as NB1 negative, however, this individual would not necessarily be NB2 positive.

4. The HLA and HNA nomenclature

In 1964, Bernard Amos of Duke University was appointed by the National Research Council to organize an international research team to identify the antigens involved in organ transplantation. Over a few years, different antigens were identified and were named 'HL-A' [81, 82], in which 'H' and 'L' were the terms used by different investigators for the antigens they had discovered, and 'A' was for the first locus identified. It was later recognized that the system was too complicated, and HL-A was changed to HLA to make place for other antigens and loci. Therefore, the

HLA is not "Human Leukocytes Antigens" nor is "Human lymphocyte Antigens," as used by the current literature. The HLA antigens are widely distributed in various tissues and not leukocyte-specific. It should also be noted that these antigens have their biological functions. and are not designed for organ transplantation that is a medical procedure. The term "HNA," proposed to be used for neutrophil-specific antigens, is a misinterpreted copy of HLA nomenclature and creates confusing issues: In the HNA nomenclature, the letter 'H' for human is wrong because these antigens are not specific for humans, and are present in various other animals, 'N' for neutrophil is wrong, because none of the antigens, HNA3, HNA4 nor HNA5, described in NHA systems are neutrophil-specific or cause ANN, AINI or neutropenia.

5. Basis of leukoagglutinin methodology

Many techniques are used for investigating neutrophils antigens [83, 84]. However, leukoagglutination is one of the most essential laboratory tests for neutrophil antibody detection. Unfortunately, some facilities avoid neutrophil agglutination testing because of the lack of reproducibility due to the use of inappropriate technology. The mechanism of neutrophil agglutination was investigated by time-lapse cinematography [85] and showed that after the addition of antibodies to neutrophil suspensions, a period of 5–10 minutes incubation at 37°C was required to activate the neutrophils. After this silent period, neutrophils develop many pseudopods and begin to move toward each other (agglutinate). This suggests that neutrophil agglutination is a chemotactic response and requires a biological environment. Damage to neutrophils during isolation, storage, presence of contamination, and centrifugation cause non-specific clumping. Another technical difficulty is the mixed agglutination that occurs when there are red cells (or red cell ghosts) in the cell preparation and red cell antibodies present. This red cell incompatibility, in the presence of complement, causes neutrophil stimulation to phagocytize the red cells, and or red cell ghosts, and form massive neutrophil aggregates. EDTA In the medium can prevent this process. Maintaining a highly viable environment for neutrophils is essential for the neutrophil agglutination test.

6. Conclusion

Neutrophil-Specific antigens are biological structures on blood neutrophils, the most frequent nucleated cells in blood circulation. Fetal-maternal incompatibility on these antigens causes neonatal neutropenia, and their autoimmunity results in neutropenia in children and adults. Also, the presence of antibodies against these antigens causes serious complications in blood transfusions, and incompatibility in bone marrow transplantation can cause graft rejection. Beyond antigenicity and immunological effects, these molecules have major roles on neutrophil functions: NA antigens connect neutrophils to antibodies to perform their phagocytic and other defense functions. The NB antigen interacts with Proteinase 3, Platelet-Endothelial Cell Adhesive Molecule 1 and other molecules to perform various neutrophil functions, including protein digestion and penetration across endothelial cells. This may be part of the mechanism of the development of serious lung injuries associated with COVID-19 infection and transfusion-related acute lung injury. It is also recognized that NB antigen expression is increased in Polycythemia Vera and in several cancer tissues. More investigation is needed to understand the significance of the appearance of NB on neoplastic tissues. This information would contribute to the understanding of the development of malignancies, their progression, and lead to the development of new approaches for their treatment.

Acknowledgements

The authors are grateful to Mr. Jeffrey Rothschild for his support of the Neurological Surgery Research Laboratory at Montefiore Medical Center. The authors also thank Shirin Lalezari and Rukmani Lekhraj for their assistance in preparation of this manuscript.

Conflict of interest

The authors declare no conflict of interest.

Author details

Parviz Lalezari[1]* and Behnaz Bayat[2]

1 Clinical Professor of Neurological Surgery, Medicine and Pathology, Montefiore Medical Center and Albert Einstein College of Medicine, New York, USA

2 Head of Research Laboratories and Deputy of Granulocyte and Immunohematology Laboratories, Institute for Clinical Hematology and Transfusion Medicine, Justus–Liebig-University, Giessen, Germany

*Address all correspondence to: lalezari2@aol.com

References

[1] Metchnikoff E. Etude sur la resorption des cellles. Annlnst Pasteur. 1899;**13**:737

[2] Doan CA. The recognition of a biologic differentiation in the white blood cells: With especial reference to blood transfusion. Journal of the American Medical Association. 1926;**86**(21):1593-1597

[3] Dausset J, Nenna A. Presence of leuko-agglutinin in the serum of a case of chronic agranulocytosis. Comptes Rendus des Seances de la Societe de Biologie et de Ses Filiales. 1952;**146**(19-20):1539-1541

[4] Brittingham TE. Immunologic studies on leukocytes. Vox Sanguinis. 1957;**2**(4):242-248

[5] Dausset J, Brecy H. Identical nature of the leucocyte antigens detectable in monozygotic twins by means of immune iso-leuco-agglutinins. Nature. 1957;**180**(4599):1430

[6] Lalezari P, Spaet TH. Studies on the genetics of leukocyte antigens. Blood. 1959;**14**(6):748-758

[7] Payne R, Hackel E. Inheritance of human leukocyte antigens. American Journal of Human Genetics. 1961;**13**:306-319

[8] Lalezari P, Nussbaum M, Gelman S, Spaet TH. Neonatal neutropenia due to maternal isoimmunization. Blood. 1960;**15**:236-243

[9] Lalezari P, Bernard GE. An isologous antigen-antibody reaction with human neutrophiles, related to neonatal neutropenia. The Journal of Clinical Investigation. 1966;**45**(11):1741-1750

[10] Lalezari P, Bernard GE. Identification of a specific leukocyte antigen: Another presumed example of 5b. Transfusion. 1965;**5**:135-142

[11] Lalezari P, Thalenfeld B, Weinstein W. The third neutrophil antigen. Histocompatibility Testing. 1970;**1970**:319-322

[12] Stroncek DF, Skubitz KM, Plachta LB, Shankar RA, Clay ME, Herman J, et al. Alloimmune neonatal neutropenia due to an antibody to the neutrophil Fc-gamma receptor III with maternal deficiency of CD16 antigen. Blood. 1991;**77**(7):1572-1580

[13] Bux J, Stein EL, Bierling P, Fromont P, Clay M, Stroncek D, et al. Characterization of a new alloantigen (SH) on the human neutrophil Fc gamma receptor IIIb. Blood. 1997;**89**(3):1027-1034

[14] Reil A, Sachs UJ, Siahanidou T, Flesch BK, Bux J. HNA-1d: A new human neutrophil antigen located on Fc gamma receptor IIIb associated with neonatal immune neutropenia. Transfusion. 2013;**53**(10):2145-2151

[15] Payne R. Neonatal neutropenia and leukoagglutinins. Pediatrics. 1964;**33**:194-204

[16] Lalezari P. Neutrophil antigens: Immunology and clinical implications. Progress in Clinical and Biological Research. 1977;**13**:209-225

[17] Lalezari P, Jiang AF, Yegen L, Santorineou M. Chronic autoimmune neutropenia due to anti-NA2 antibody. The New England Journal of Medicine. 1975;**293**(15):744-747

[18] Lalezari P, Khorshidi M, Petrosova M. Autoimmune neutropenia of infancy. The Journal of Pediatrics. 1986;**109**(5):764-769

[19] Verheugt FW, von dem Borne AE, van Noord-Bokhorst JC, van Elven EH, Engelfriet CP. Serological, immunochemical and

immuoncytological properties of granulocyte antibodies. Vox Sanguinis. 1978;**35**(5):294-303

[20] Claas FH, Langerak J, Sabbe LJ, van Rood JJ. NE1: A new neutrophil specific antigen. Tissue Antigens. 1979;**13**(2): 129-134

[21] Lalezari P. Autoimmune Neutropenia. The Autoimmune Diseases. Amsterdam, Netherlands: Elsevier; 2014. pp. 677-683

[22] Lalezari P. Neutrophil-specific antigens, immunobiology, and implications in transfusion medicine and blood disorders. Transfusion. 2017;**57**(9):2066-2073

[23] Huizinga TW, Kuijpers RW, Kleijer M, Schulpen TW, Cuypers HT, Roos D, et al. Maternal genomic neutrophil Fc RIII deficiency leading to neonatal isoimmune neutropenia. Blood. 1990;**76**(10):1927-1932

[24] Clark MR, Liu L, Clarkson SB, Ory PA, Goldstein IM. An abnormality of the gene that encodes neutrophil Fc receptor III in a patient with systemic lupus erythematosus. The Journal of Clinical Investigation. 1990;**86**(1):341-346

[25] Huizinga TW, van der Schoot CE, Jost C, Klaassen R, Kleijer M, von dem Borne AE, et al. The PI-linked receptor FcRIII is released on stimulation of neutrophils. Nature. 1988;**333**(6174):667-669

[26] Selvaraj P, Rosse WF, Silber R, Springer TA. The major Fc receptor in blood has a phosphatidylinositol anchor and is deficient in paroxysmal nocturnal haemoglobinuria. Nature. 1988;**333**(6173):565-567

[27] Werner G, Von dem Borne AK, Bos M, Tromp J, Van der Plas-Van DC, Visser F, et al. Localization of the Human NA1 Alloantigen on Neutrophil Fc-γ-Receptors. Leukocyte typing II. Berlin, Germany: Springer; 1986. pp. 109-121

[28] Lu X, Machiesky LA, De Mel N, Du Q, Xu W, Washabaugh M, et al. Characterization of IgG1 Fc Deamidation at asparagine 325 and its impact on antibody-dependent cell-mediated cytotoxicity and FcgammaRIIIa binding. Scientific Reports. 2020;**10**(1):383

[29] van Sorge NM, van der Pol WL, van de Winkel JG. FcgammaR polymorphisms: Implications for function, disease susceptibility and immunotherapy. Tissue Antigens. 2003;**61**(3):189-202

[30] Paulick MG, Forstner MB, Groves JT, Bertozzi CR. A chemical approach to unraveling the biological function of the glycosylphosphatidylinositol anchor. Proceedings of the National Academy of Sciences of the United States of America. 2007;**104**(51):20332-20337

[31] van der Pol W, van de Winkel JG. IgG receptor polymorphisms: Risk factors for disease. Immunogenetics. 1998;**48**(3):222-232

[32] Ravetch JV, Perussia B. Alternative membrane forms of Fc gamma RIII(CD16) on human natural killer cells and neutrophils. Cell type-specific expression of two genes that differ in single nucleotide substitutions. The Journal of Experimental Medicine. 1989;**170**(2):481-497

[33] Sachs UJ, Radke C, Bein G, Grabowski C, Simtong P, Bux J, et al. Primary structure of human neutrophil antigens 1a and 1b. Transfusion. 2020;**60**(4):815-821

[34] Ory PA, Goldstein IM, Kwoh EE, Clarkson SB. Characterization of polymorphic forms of Fc receptor III on human neutrophils. The Journal of Clinical Investigation. 1989;**83**(5): 1676-1681

[35] Fromont P, Bettaieb A, Skouri H, Floch C, Poulet E, Duedari N, et al. Frequency of the polymorphonuclear neutrophil Fc gamma receptor III deficiency in the French population and its involvement in the development of neonatal alloimmune neutropenia. Blood. 1992;**79**(8):2131-2134

[36] Koene HR, Kleijer M, Roos D, de Haas M, Von dem Borne AE. Fc gamma RIIIB gene duplication: Evidence for presence and expression of three distinct Fc gamma RIIIB genes in NA(1+,2+)SH(+) individuals. Blood. 1998;**91**(2):673-679

[37] Treffers LW, van Houdt M, Bruggeman CW, Heineke MH, Zhao XW, van der Heijden J, et al. FcgammaRIIIb restricts antibody-dependent destruction of Cancer cells by human neutrophils. Frontiers in Immunology. 2018;**9**:3124

[38] de Haas M, Kleijer M, van Zwieten R, Roos D, von dem Borne AE. Neutrophil Fc gamma RIIIb deficiency, nature, and clinical consequences: A study of 21 individuals from 14 families. Blood. 1995;**86**(6):2403-2413

[39] Nath SK, Han S, Kim-Howard X, Kelly JA, Viswanathan P, Gilkeson GS, et al. A nonsynonymous functional variant in integrin-alpha(M) (encoded by ITGAM) is associated with systemic lupus erythematosus. Nature Genetics. 2008;**40**(2):152-154

[40] Tosi MF, Zakem H. Surface expression of Fc gamma receptor III (CD16) on chemoattractant-stimulated neutrophils is determined by both surface shedding and translocation from intracellular storage compartments. The Journal of Clinical Investigation. 1992;**90**(2):462-470

[41] van de Winkel JG, Capel PJ. Human IgG Fc receptor heterogeneity: Molecular aspects and clinical implications. Immunology Today. 1993;**14**(5):215-221

[42] Bux J, Behrens G, Leist M, Mueller-Eckhardt C. Evidence that the granulocyte-specific antigen NC1 is identical with NA2. Vox Sanguinis. 1995;**68**(1):46-49

[43] Huizinga TW, de Haas M, Kleijer M, Nuijens JH, Roos D, von dem Borne AE. Soluble Fc gamma receptor III in human plasma originates from release by neutrophils. The Journal of Clinical Investigation. 1990;**86**(2):416-423

[44] Lalezari P, Murphy GB, Allen FH Jr. NB1, a new neutrophil-specific antigen involved in the pathogenesis of neonatal neutropenia. The Journal of Clinical Investigation. 1971;**50**(5):1108-1115

[45] Stroncek DF, Caruccio L, Bettinotti M. CD177: A member of the Ly-6 gene superfamily involved with neutrophil proliferation and polycythemia vera. Journal of Translational Medicine. 2004;**2**(1):8

[46] Caruccio L, Walkovich K, Bettinotti M, Schuller R, Stroncek D. CD177 polymorphisms: Correlation between high-frequency single nucleotide polymorphisms and neutrophil surface protein expression. Transfusion. 2004;**44**(1):77-82

[47] Stroncek DF, Shankar R, Litz C, Clement L. The expression of the NB1 antigen on myeloid precursors and neutrophils from children and umbilical cords. Transfusion Medicine. 1998;**8**(2):119-123

[48] Korkmaz B, Kuhl A, Bayat B, Santoso S, Jenne DE. A hydrophobic patch on proteinase 3, the target of autoantibodies in Wegener granulomatosis, mediates membrane binding via NB1 receptors. The Journal of Biological Chemistry. 2008;**283**(51):35976-35982

[49] Stroncek DF, Skubitz KM, McCullough JJ. Biochemical characterization of the

neutrophil-specific antigen NB1. Blood. 1990;**75**(3):744-755

[50] Skubitz KM, Stroncek DF, Sun B. Neutrophil-specific antigen NB1 is anchored via a glycosyl-phosphatidylinositol linkage. Journal of Leukocyte Biology. 1991;**49**(2):163-171

[51] Goldschmeding R, van Dalen CM, Faber N, Calafat J, Huizinga TW, van der Schoot CE, et al. Further characterization of the NB 1 antigen as a variably expressed 56-62 kD GPI-linked glycoprotein of plasma membranes and specific granules of neutrophils. British Journal of Haematology. 1992;**81**(3):336-345

[52] Wu Z, Liang R, Ohnesorg T, Cho V, Lam W, Abhayaratna WP, et al. Heterogeneity of human neutrophil CD177 expression results from CD177P1 Pseudogene conversion. PLoS Genetics. 2016;**12**(5):e1006067

[53] Caruccio L, Bettinotti M, Director-Myska AE, Arthur DC, Stroncek D. The gene overexpressed in polycythemia rubra vera, PRV-1, and the gene encoding a neutrophil alloantigen, NB1, are alleles of a single gene, CD177, in chromosome band 19q13.31. Transfusion. 2006;**46**(3):441-447

[54] Temerinac S, Klippel S, Strunck E, Roder S, Lubbert M, Lange W, et al. Cloning of PRV-1, a novel member of the uPAR receptor superfamily, which is overexpressed in polycythemia rubra vera. Blood. 2000;**95**(8):2569-2576

[55] Kissel K, Santoso S, Hofmann C, Stroncek D, Bux J. Molecular basis of the neutrophil glycoprotein NB1 (CD177) involved in the pathogenesis of immune neutropenias and transfusion reactions. European Journal of Immunology. 2001;**31**(5):1301-1309

[56] Dittmar K, Lim JB, Caruccio L, Bettinotti M, Stroncek D. Assessment of the relative number of copies of the gene encoding human neutrophil antigen-2a(HNA-2a), CD177, and a homologous pseudogene by quantitative real-time PCR. Immunohematology. 2003;**19**(4):122-126

[57] Li Y, Mair DC, Schuller RM, Li L, Wu J. Genetic mechanism of human neutrophil antigen 2 deficiency and expression variations. PLoS Genetics. 2015;**11**(5):e1005255

[58] Bayat B, Bein G, Sachs UJ. A sequence-specific polymerase chain reaction method for HNA-2 genotyping: Homozygous c.843A>T mutation predicts the absence of CD177. Transfusion. 2016;**56**(8):2127-2132

[59] Flesch BK, Reil A, Nogues N, Canals C, Bugert P, Schulze TJ, et al. Multicenter study on differential human neutrophil antigen 2 expression and underlying molecular mechanisms. Transfusion Medicine and Hemotherapy. 2020;**47**(5):385-395

[60] Wu J, Li Y, Schuller RM, Li L, Litmeyer AS, Bein G, et al. The nonconservative CD177 single-nucleotide polymorphism c.1291G>A is a genetic determinant for human neutrophil antigen-2 atypical/low expression and deficiency. Transfusion. 2019;**59**(5):1836-1842

[61] Eulenberg-Gustavus C, Bahring S, Maass PG, Luft FC, Kettritz R. Gene silencing and a novel monoallelic expression pattern in distinct CD177 neutrophil subsets. The Journal of Experimental Medicine. 2017; **214**(7):2089-2101

[62] Jerke U, Marino SF, Daumke O, Kettritz R. Characterization of the CD177 interaction with the ANCA antigen proteinase 3. Scientific Reports. 2017;7:43328

[63] Traum A, Hofmann C, Haas S, Schmidt S, Bein G, Sachs UJ, et al.

Characterization of CD177-reactive iso- and auto-antibodies. Transfusion. 2021;**61**(6):1916-1922

[64] Bayat B, Werth S, Sachs UJ, Newman DK, Newman PJ, Santoso S. Neutrophil transmigration mediated by the neutrophil-specific antigen CD177 is influenced by the endothelial S536N dimorphism of platelet endothelial cell adhesion molecule-1. Journal of Immunology. 2010;**184**(7):3889-3896

[65] Sachs UJ, Hattar K, Weissmann N, Bohle RM, Weiss T, Sibelius U, et al. Antibody-induced neutrophil activation as a trigger for transfusion-related acute lung injury in an ex vivo rat lung model. Blood. 2006;**107**(3):1217-1219

[66] Bai M, Grieshaber-Bouyer R, Wang J, Schmider AB, Wilson ZS, Zeng L, et al. CD177 modulates human neutrophil migration through activation-mediated integrin and chemoreceptor regulation. Blood. 2017;**130**(19):2092-2100

[67] Kuckleburg CJ, Tilkens SB, Santoso S, Newman PJ. Proteinase 3 contributes to transendothelial migration of NB1-positive neutrophils. Journal of Immunology. 2012;**188**(5):2419-2426

[68] Xie Q, Klesney-Tait J, Keck K, Parlet C, Borcherding N, Kolb R, et al. Characterization of a novel mouse model with genetic deletion of CD177. Protein & Cell. 2015;**6**(2):117-126

[69] Hu N, Westra J, Kallenberg CG. Membrane-bound proteinase 3 and its receptors: Relevance for the pathogenesis of Wegener's Granulomatosis. Autoimmunity Reviews. 2009;**8**(6):510-514

[70] Zhou G, Yu L, Fang L, Yang W, Yu T, Miao Y, et al. CD177(+) neutrophils as functionally activated neutrophils negatively regulate IBD. Gut. 2018;**67**(6):1052-1063

[71] Alayed K, Meyerson JB, Osei ES, Blidaru G, Schlegelmilch J, Johnson M, et al. CD177 enhances the detection of myelodysplastic syndrome by flow Cytometry. American Journal of Clinical Pathology. 2020;**153**(4):554-565

[72] Huang YH, Lo MH, Cai XY, Liu SF, Kuo HC. Increase expression of CD177 in Kawasaki disease. Pediatric Rheumatology Online Journal. 2019;**17**(1):13

[73] Kluz PN, Kolb R, Xie Q, Borcherding N, Liu Q, Luo Y, et al. Cancer cell-intrinsic function of CD177 in attenuating beta-catenin signaling. Oncogene. 2020;**39**(14):2877-2889

[74] Bayat B, Nielsen KR, Bein G, Traum A, Burg-Roderfeld M, Sachs UJ. Transfusion of target antigens to preimmunized recipients: A new mechanism in transfusion-related acute lung injury. Blood Advances. 2021;**5**(20):3975-3985

[75] Nicolai L, Leunig A, Brambs S, Kaiser R, Weinberger T, Weigand M, et al. Immunothrombotic dysregulation in COVID-19 pneumonia is associated with respiratory failure and coagulopathy. Circulation. 2020;**142**(12):1176-1189

[76] Levy Y, Wiedemann A, Hejblum BP, Durand M, Lefebvre C, Surenaud M, et al. CD177, a specific marker of neutrophil activation, is associated with coronavirus disease 2019 severity and death. iScience. 2021;**24**(7):102711

[77] Stroncek DF, Shankar RA, Plachta LB, Clay ME, Dalmasso AP. Polyclonal antibodies against the NB1-bearing 58- to 64-kDa glycoprotein of human neutrophils do not identify an NB2-bearing molecule. Transfusion. 1993;**33**(5):399-404

[78] Lalezari P, Petrosova M, Jiang AF. NB2, an allele of NB1 neutrophil specific antigen: Relationship to 9a (abstract). Transfusion. 1982;**22**:433

[79] Huang ST, Lin J, McGowan E, Lalezari P. NB2, a new allele of NB1 antigen, involved in febrile transfusion reaction (abstract). Transfusion. 1982;**22**:426

[80] Van Buren NL, Stroncek DF, Clay ME, McCullough J, Dalmasso AP. Transfusion-related acute lung injury caused by an NB2 granulocyte-specific antibody in a patient with thrombotic thrombocytopenic purpura. Transfusion. 1990;**30**(1):42-45

[81] Amos DB. Human histocompatibility locus HL-A. Science. 1968;**159**(3815):659-660

[82] Amos DB. Why "HLA". ASHI: The First 25 Years, 1974-1999. ASHI (American Society of Histocompatibility and Immunology); 1999

[83] McCullough J. Granulocyte serology: A clinical and laboratory guide. Chicago: ASCP Press (American Society clinical Pathologists); 1988

[84] Fung YL, Minchinton RM, Fraser JF. Neutrophil antibody diagnostics and screening: Review of the classical versus the emerging. Vox Sanguinis. 2011;**101**(4):282-290

[85] Lalezari P. Neutrophil antigens: Immunology and clinical implications. In: Greenwalt TJ, Jamieson GA, editors. Granulocyte: Function and Clinical Utilization. New York: Alan R. Liss; 1977. pp. 209-225

Chapter 8

The ABO Blood Group System and *Plasmodium falciparum* (*Pf*) Infection in Three Ethnic Groups Living in the Stable and Seasonal Malaria Transmission Areas of Burkina Faso (BF)

Edith Christiane Bougouma, Alphonse Ouedraogo and Sodiomon Bienvenu Sirima

Abstract

Genetic factors, including red blood cell polymorphisms, influence the severity of disease due to infection with *Plasmodium falciparum (Pf)*. Studies show that these genetic factors associated with malaria susceptibility or resistance vary geographically, ethnically, and racially. We performed cross-sectional surveys in population living in rural villages from three ethnic groups. The blood group (BG) was determined genetically using two polymorphisms (rs8176719 and rs8176746). Out of 548 participants, 29.7% were Mossi, 38.2% were Fulani, and 32.1% were Rimaibe. The distribution of BG was, respectively, A: 25.5%, B: 26.6%, AB: 7.3%, and O: 40.5%. BG O was not only the common blood type overall, but was higher in Fulani (52.6%) than others. Fulani was associated with a reduced risk of infection and lower parasite densities than sympatric populations. The subjects with non-O blood were less susceptible to malaria infection. An association between ethnicity and malaria infection during the high transmission season as well as an association between the non-O blood group and malaria infections according to ethnicity was found. This was also true when ethnic groups were considered separately. Our results have demonstrated that the Fulani are not only less susceptible to *Pf* malaria infection, but when infected have lower parasite densities. Individuals with non-O blood are at lower risk of infection.

Keywords: ABO blood group system, malaria, *Plasmodium falciparum*, ethnic groups, Burkina Faso

1. Introduction

Malaria is one of the prevalent infectious diseases globally. In malaria-endemic areas, a significant proportion of children harbor parasites without presenting signs of clinical malaria and are considered asymptomatic cases [1]. Moreover, in communities where children are repeatedly infected with malaria, one can question why some children die while others do not. Variant-specific immunity may help explain chronic low-grade malaria infection without clinical symptoms [2]. A better knowledge of the polymorphic host genes associated with resistance to clinical malaria and/or with high parasite densities might provide new insights into disease mechanisms, and suggest new approaches for prophylactic or therapeutic interventions [3, 4].

Plasmodium falciparum (*Pf*) has been called "the strongest known force for evolutionary selection in the recent history of the human genome [5, 6]. Indeed, the high mortality associated with *Pf* has given it the capacity to select emerging polymorphisms as rapidly as can be witnessed in evolutionary time [7, 8]. Several human genetic factors including red blood cell polymorphisms influence the severity of disease due to infection with *Pf*. To date, there is a paucity of information concerning the role of these host genetic factors (ABO blood group, sickle-cell trait, G6PD deficiency) in asymptomatic malaria, characterized by low-grade *Pf* infection and absence of fever/overt symptoms.

The ABO blood groups consist of A, B, and H carbohydrate antigens which can regulate protein activities during infection and antibodies against these antigens [9–11]. Several studies did try to establish an association between severe malaria and the ABO-blood group type [11–13], with some reporting significant associations on infection status and a particular ABO blood group [11, 13].

The relationship between ABO and malaria was first suggested more than 40 years ago [14]. Few studies have corroborated this hypothesis [8, 11, 14, 15], but some studies showed a weak association [2]. In India, for example, a significantly lower frequency of *Pf* infection was observed among individuals with blood groups A and O [9]. Thus, the selection pressures defining the ABO distributions remain uncertain with contrasting results between studies.

This study aimed to investigate the association between the ABO blood group and malaria susceptibility among Fulani compared to other sympatric ethnic groups living in Burkina Faso.

We offer four perspectives in support of this investigation:

- The prevalence of the ABO blood group, in both children and adults residing in the rural areas of Burkina Faso;

- The current distribution of ABO groups and *Pf* infection prevalence;

- Clinical outcomes during *Pf* infection; and - Relationship between the ABO blood group and the prevalence of asymptomatic *Pf* infection.

2. Brief review of malaria infection ABO blood groups

For a long time ago, the ABO blood group system has been suggested to be associated with infectious diseases including malaria. ABO blood groups appear to protect against malaria, through loss of function due to a defective allele: the O phenotype of the ABO system seems to confer protection against *Pf* infection (see **Table 1**).

The human ABO gene consists of seven exons that are more than 18 kb in length and genomic analysis has found over 70 alleles at this locus, suggesting that it is one of the polymorphic genes in humans [1]. The three main antigenic classes (A, B, and O) are all comprised of numerous alleles in both coding and non-coding sequences [1].

The O alleles share a one-nucleotide deletion in codon 87 of exon 6 resulting in a frameshift mutation and premature termination of the polypeptide [2, 3]. O alleles are the most common of the three allelic classes (about 0.6 worldwide) and have frequencies between 0.3 and 0.7 in most populations. The A alleles generally have

Gene (chromosome)	Protein	Mutation	Number of mutations	Variant	Reported Genetic Associations with Malaria	Mechanistic hypotheses proposed protective mechanism	Reference	Distribution	High-low Frequency
ABO (9q34) O	Glycosyl transferase enzyme	Deletion of nucleotide 261 in exon 6	major	ABO single nucleotide deletion (rs8176719)-Blood group O	O alleles protect against uncomplicated malaria and severe malaria.	Reduced *P. falciparum* rosetting	Rowe et al. (2007), Rowe et al. (1995), Udomsangpetch et al. (1993), MalariaGEN (2014), Ndila et al. (2018), Tishkoff SA, et al. (2004). Silvia N. Kariuki	South America, Africa, Western Europe	Near 100% in native South Americans to about 0.3% in some Asians

Table 1.
How the ABO blood group system affect susceptibility and resistance to P. falciparum *malaria.*

frequencies between 0.2 and 0.3, and the B alleles have frequencies between 0.1 and 0.2 [4]. There is substantial evidence supporting the importance of allele O for protecting against malaria, based primarily on the consistency between the worldwide distributions of ABO variants and the historical presence of malaria. Uneke (2007) showed that the O allele, which is more frequent in malaria-endemic regions, is associated with greater resistance to malaria, in contrast to the A and B genotypes, which are less resistant [5]. Fry et al. [6] corroborated this finding through a study in three African populations showing a strong association of O individuals with resistance to severe malaria and a recessive effect in AO and BO individuals. Consequently, AO and BO would have the same susceptibility or sensitivity to malaria as AA and BB individuals; with AB genotype individuals being the most sensitive. The O allele is thought to protect against severe malaria through a mechanism of reduced reinitiation (*i.e.,* spontaneous binding of infected erythrocytes to uninfected erythrocytes) [7].

However, although the major O alleles share a one-nucleotide deletion, they differ in the number of nucleotide substitutions in both exons and introns. The O human alleles, although different from the O chimpanzee allele [8], are much older [1]. The most common O alleles, O01 and O02, are the result of separate mutations and are 1.15 and 2.5 million years old, respectively. In other words, assuming that the O allele protects from malaria, this protection may be a result of selection originally favoring O alleles for some other reason.

3. Method

3.1 Study area

The study was conducted in four shrub-savanna villages located northeast (Barkoumbilen and Barkoundouba) and east (Bassy and Zanga) of Ouagadougou, the capital city of Burkina Faso (see **Figure 1**).

In the northeast zone, the Mossi and Rimaibè communities inhabit the village of Barkoumbilen, while the Fulani and Rimaibè inhabit the village of Barkoundouba. The Fulani and Rimaibe are present in both villages, so the Rimaibe represent an optimal internal control for the study. The two villages are 5 km apart. In the eastern zone, Mossi and Fulani communities independently inhabit two villages, Zanga and Bassy, about 1(one) km apart.

Malaria transmission in the study areas is hyperendemic and seasonal, with a rainy season from June to October. The entomological inoculation rate (EIR) is estimated to be around 200–300 infecting bites per person per year and is comparable between villages [6]. The main malaria vectors are *Anopheles gambiae* ss, *An. arabiensis* and *An. funestus*. *Plasmodium falciparum* (*Pf*) is responsible for more than 90% of malaria infections.

3.2 Study participants and study period

The population of the study area is predominantly of the Mossi ethnic group, followed by the widespread Fulani, who are generally closely associated with the Rimaibe (a hybrid of the two in terms of customs and even genetics). All three populations live in similar houses. However, the Fulani group, while adopting the habits of the sedentary Mossi populations, has retained their pastoral activities so that their habitat is characterized by the presence of cattle herds.

The total population of the four villages was obtained from a general census and enrolled in a demographic monitoring process. Except for children under 6 months

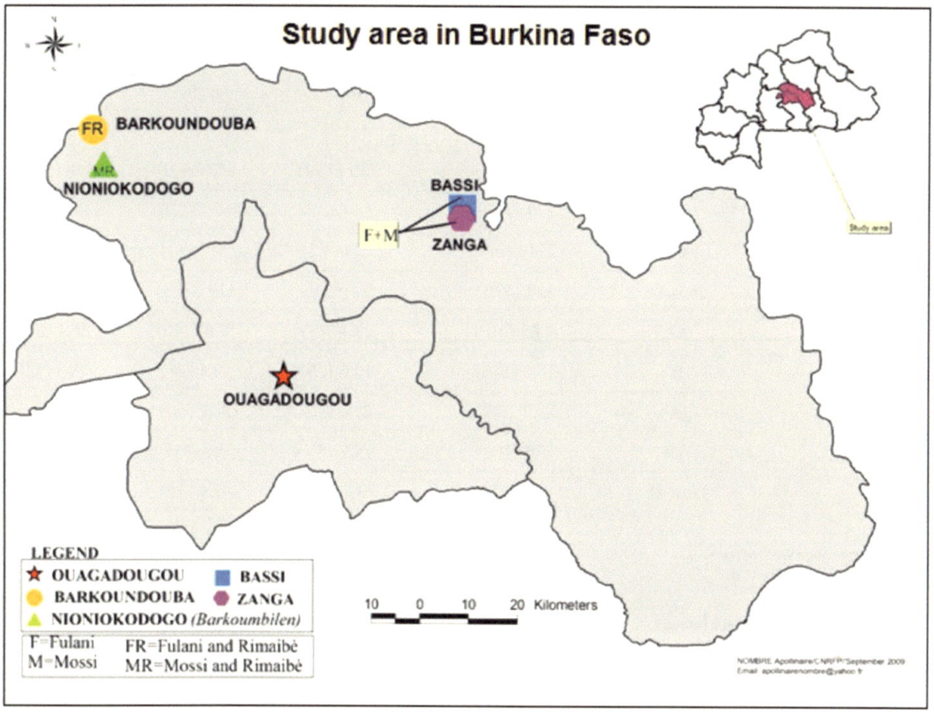

Figure 1.
Study site in Burkina. The study was carried out in four rural villages of shrubby savannah areas of Burkina Faso: northeast (Nioniokodogo and Barkoundouba), and east (Bassy and Zanga) of the capital town Ouagadougou. Compounds of the various ethnic groups are represented as capital letters. Inset/smaller map represents the study site relative to the whole country.

of age, the entire population was eligible for the study and was characterized/designated into two groups: participants aged 0.6 to 17 years and participants aged 18 years and older.

The study was conducted for two consecutive years and consisted of five cross-sectional surveys (see **Table 2**).

i. Four surveys during the high malaria transmission seasons: in the middle of August 2007 and 2008; and in the middle of November/December 2007 and 2008.

ii. One survey in the middle of the low transmission/dry season: March 2008.

3.3 Data collection

Community sensitization meetings were held during which information or explanations about the context of the study, its objectives, its methodological approach, and the associated ethical issues were provided. The signing of written informed consent was a requirement for all study participants, including the guardians of minor children.

Following consent, a clinical team of physicians and nurses examined all participants for clinical signs of malaria, measured weight, and axillary body temperature. Suspected malaria cases ($T \geq 37.5°C$) were treated with artemether and lumefantrine.

During the first cross-sectional survey, a two (2) ml venous blood sample was collected from all participants in EDTA tubes.

Thick and thin blood smears were prepared and the hemoglobin level was measured by the HemoCue technique.

		Total subject (n)	Grouped by ethnicity		
			Mossi n (%)	Fulani n (%)	Rimaibe n (%)
Number/subjects		559	163	209	176
Age group (years)	**0.6–17**	380 (68.1)	120 (73.2)	147 (69.0)	113 (62.4)
	>18	178 (31.9)	44 (26.8)	66 (31.0)	68 (37.6)
Gender	**Male**	224 (40.9)	72 (44.2)	90 (43.1)	62 (35.2)
	Female	324 (59.1)	91 (55.8)	119 (56.9)	114 (64.8)
Blood group	**O**	40.5 (222)	29.4 (48)	52.6 (110)	36.4 (64)
	B	26.6 (146)	40.5 (66)	13.4 (28)	29.5 (52)
	A	25.5 (140)	25.2 (41)	28.7 (60)	22.2 (39)
	AB	7.3 (40)	4.9 (8)	5.3 (11)	52.5 (21)
	Non-O	59.2 (326)	70.6 (115)	47.4 (99)	63.6 (112)

Table 2.
Population demographic.

3.4 Laboratory methods

3.4.1 Sample collection

For each subject in this study, physical examination and capillary blood samples on slides and filter papers were collected. Slides were used for the diagnosis of malaria infection and the venous sampling was for the blood group characterization.

3.4.2 Blood slides

Blood slides were stained with Giemsa for microscope identification of the *Plasmodium* species and determination of parasite density.

3.4.3 Malaria parasite diagnosis by microscopy

Blood films were air-dried, the thin films fixed with methanol, and the slide stained with Giemsa. The slides were read by highly experienced laboratory technicians according to the site SOPs. Briefly, 100 high power fields were examined, and the number of malaria parasites/each species and stage was recorded. The number of parasites per microliter of blood was calculated assuming 200 white blood cells per high power field and a fixed white blood cell count of 8000/μL. A slide was considered negative if no parasites were found after the 100 HPF examination. Two independent microscopists read each slide and in case of discrepancy between the two readers, in terms of species, presence or absence of malaria parasites, or if the parasite densities differed by more than 30%, the slide was re-examined by a third laboratory technician. The arithmetic mean of the two closest readings was used as the final value of the parasite density. If there was no agreement after the third reading, the arithmetic mean of the three parasite densities was used.

3.4.4 DNA samples

DNA extraction from the buffy coat (about 1.5 ml in volume) was performed using Nucleon BACC2 Kits. Successful DNA extraction was checked on a 1% agarose gel stained with ethidium bromide. A total of 2235 DNA samples (about 50 μl in volume) were collected: 825 from Mossi, 877 from Fulani, and 533 from Rimaibè individuals. Samples were stored in screw cap tubes labeled with the study code. Samples were kept at −20°C in cryo boxes. The DNA samples were shipped to the WTCHG in Oxford, UK, in May 2008. The DNA samples of study participants were genotyped at the WTCHG in Oxford, UK.

3.5 Data management

Demographic variables collected during the census of the four villages include for each person; age, gender, father and mother ethnicity, village, compound, family members and sequential number within the family, and census code. Assignment of ethnicity has been performed with the assistance of local guides.

The census data file was anonymized (name identification has been removed for privacy protection) according to the study protocol which was approved by the ethical committee of Burkina Faso. Clinical data were matched to the census data with study enrolment codes. Case report forms (CRF) for cross-sectional surveys were verified by the supervisor of the clinical study before data entry. Similarly, parasitological data were verified by the lab supervisor before data entry. Data entry was conducted by a team of computer engineers at Ouagadougou, Burkina Faso, and data files were validated by the database manager.

3.6 Statistical analysis

Data were entered into Microsoft Access. A Chi-square test was used to assess the difference between frequencies/associations between blood groups and *Pf* cases. ANOVA was used to test the difference between mean parasitemia.

Two-sided p values were reported, with differences considered significant at $p \leq 0.05$. All analyses were carried out with the statistical software R 2.10.

3.7 Clearance from National Ethics Committee

The National Ethics Committee of the Ministry of Health (in Burkina Faso) granted ethical clearance. The study was conducted in compliance with the International Conference on Harmonization, Good Clinical Practices, the Declaration of Helsinki, and applicable Burkina Faso regulatory requirements. Individual written informed consent was obtained from each participant, participant's parents, or legally acceptable representative.

4. Results

This study aimed to investigate *Pf* infection in Burkina Faso in the different ethnic groups: Fulani, Rimaibe, and Mossi according to the ABO group.

4.1 Characteristics of the study participants

A total of 548 subjects (380 children and 178 adults) were included in this study: Mossi, n = 163 (29.7%), Fulani, n = 209 (38.2%), and Rimaibe, n = 176 (32.1%). **Table 2** shows baseline demographic characteristics according to ethnicity. We did not find any significant difference among the three ethnic groups. The ABO blood group analysis in all children revealed that O antigen: 40.5% (222/548) was the most predominant, followed by A: 25.5% (140/548), B: 26.6% (146/548), and AB: 7.3% (40/548) antigens.

Blood group O was not only the commonest blood type overall, but was higher in the Fulani (n = 110 (52.6%)) than Mossi (n = 48 (29.4%)) and Rimaibe (n = 64 (36.4%)).

4.2 Malaria infection

The malariometric indices for each cross-sectional survey are summarized in **Table 3**. The prevalence of *Pf* infection decreased from 51.6% during the high transmission season to 18.4% in the low season; the prevalence of clinical malaria also decreased significantly from 9.5% to 0.7%. The geometric mean parasite density according to the season (high transmission season and low transmission season) was represented in **Table 3**.

4.3 Malaria infection and ethnicity

The study showed that during the period of high transmission, there was an association between ethnicity and malaria infection (p = 0.039). Subjects from the Fulani ethnic group were associated with a reduced risk of *Pf* infection (0.0001 and p = 0.02 for Fulani vs. Mossi/Rimaibe respectively during the start and end of the high transmission season). As a result, the Fulani ethnic group had lower parasite densities than the sympatric populations during the high transmission season: The mean density by ethnicity was 374 (251–557), respectively; 865 (598–1252); 483 (331–708) parasites/µl for Fulani, Mossi & Rimaibe (**Table 4**).

Malaria indices	*survey number	Total	Ethnic Group			P-value		
			Mossi	Fulani	Rimaibe	F vs. M vs. R	F VS M	F vs. R
Prevalence of Pf malaria Infection % (number positive/ total)	1	283 (51.6)	98 (60.1)	91 (43.5)	59 (52.7)	0.08	0.025	0.30
	2	236 (43.1)	90 (55.2)	54 (25.8)	54 (48.2)	0.002	0.003	0.00
	3	101 (18.4)	33 (20.24)	25 (12.0)	26 (23.2)	0.55	0.56	0.50
	4	272 (49.6)	103 (63.2)	69 (33.0)	65 (58.0)	0.0001	0.016	0.0001
	5	215 (39.2)	87 (53.4)	52 (24.9)	47 (42.0)	0.0056	0.0012	0.25
Geometric mean of parasite density (p/µl)	1	550 [441–687]	865 [598–1252]	374 [251–557]	483 [331–708]	0.021	0.001	0.316
	2	313 [250–390]	330 [228–478]	297 [192–460]	305 [211–441]	0.932	0.634	0.050
	3	393 [289–536]	230 [138–382]	391 [195–780]	572 [360–909]	0.044	0.912	0.332
	4	264 [697–1040]	685 [491–957]	878 [578–1332]	1030 [571–1414]	0.253	0.036	0.51
	5	720 [563–920]	791 [535–1167]	523 [327–836]	805 [520–1249]	0.345	0.171	0.193

Malaria indices	*survey number	Total	Ethnic Group			P-value		
			Mossi	Fulani	Rimaibe	F vs. M vs. R	F VS M	F vs. R
Prevalence of clinical Pf cases % (number positive/total)	1	52 (9.5)	16 (9.8)	18 (8.6)	18 (10.2)	0.98	0.88	0.78
	2	13 (2.4)	8 (4.9)	4 (1.9)	1 (0.1)	—	—	—
	3	4 (0.7)	0 (0.0)	1 (0.5)	3 (1.7)	—	—	—
	4	31 (5.7)	11 (6.7)	9 (4.3)	11 (6.3)	—	—	—
	5	24 (4.4)	13 (8.0)	5 (2.4)	6 (3.4)	—	—	—

**Survey number.*
1: First survey, the middle of high malaria transmission season (August 2008).
2: Second survey, the end of high malaria transmission season (Nov 2008).
3: Third survey, the middle of the dry low transmission season (March 2008).
4: The fourth survey, the start of the high transmission season (July 2008).
5: Fifth survey, the end of the high transmission season (November/December 2008).

Table 3.
Malariometric indices according to the ethnicity group/survey.

Malaria indices	Survey number		Total	Ethnic group			P-value		
				Mossi	Fulani	Rimaibe	F vs. M vs. R	F VS M	F vs. R
High transmission season (survey 1, August 2008)									
Prevalence of Pf malaria Infection % (number)	O		98 (60.1)	27 (56.3)	46 (50.9)	29 (45.3)	0.724	0.64	0.60
	Non-O		65 (39.9)	71 (61.7)	45 (45.5)	59 (52.7)	0.172	0.05	0.41
	Non-O	A	4.7 (5)	27 (38.0)	31 (33.7)	22 (23.4)	0.55	0.70	0.44
		B	4.7 (5)	41 (57.7)	6 (6.6)	24 (22.5)	0.001	0.02	0.003
		AB	9.3 (10)	3 (38.0)	8 (8.8)	13 (13.8)	—	—	—
Geometric mean of parasite density (p/µl) [positive/total]	O		521 [359–755]	1020 [486–2146]	434 [237–767]	373 [202–689]	0.07	0.07	0.71
	Non-O		569 [431–753]	811 [525–1253]	323 [188–555]	563 [343–923]	0.01	0.025	0.19
	Non-O	A	568 [382–845]	1005 [489–2068]	316 [167–599]	595 [295–1205]	0.04	0.012	0.173
		B	535 [337–851]	691 [393–1218]	136 [23–822]	494 [194–1260]	0.16	0.04	0.19
		AB	680 [226–1561]	928 [–]	664 [128–3437]	642 [204–2018]	0.90	0.50	0.92

Malaria indices	Survey number		Total	Ethnic group			P-value		
				Mossi	Fulani	Rimaibe	F vs. M vs. R	F VS M	F vs. R
Prevalence of clinical Pf cases % (number positive/total)	O		20 (9.0)	4 (8.3)	9 (8.2)	7 (10.9)	—	—	—
	Non-O		32 (9.8)	12 (10.4)	9 (9.1)	11 (9.8)	—	—	—
	Non-O	A	7 (5.0)	3 (7.3)	3 (5.0)	1 (2.6)	—	—	—
		B	19 (13.0)	8 (50.0)	5 (27.8)	4 (22.2)	—	—	—
		AB	6 (15.0)	1 (12.5)	1 (9.1)	46 (19.6)	—	—	—
Low transmission season (survey 3, Marsh 2008)									
Prevalence of Pf malaria Infection % (number)	O		39 (17.6)	10 (20.8)	12 (10.9)	17 (26.6)	0.38	0.86	0.36
	Non-O		62 (19.0)	23 (20.0)	13 (13.1)	26 (23.2)	0.85	0.98	0.88
	Non-O	A	27 (19.3)	11 (26.8)	8 (13.3)	8 (20.5)	0.72	0.83	1.00
		B	26 (17.8)	11 (16.7)	3 (10.7)	12 (23.1)	—	—	—
		AB	9 (25.5)	1 (12.5)	2 (18.2)	6 (28.6)	—	—	—
Geometric mean of parasite density (p/µl) [positive/ total]	O		345 [207–575]	185 [75–454]	434 [237–767]	483 [203–1143]	0.18	0.34	0.57
	Non-O		393 [288–536]	253 [130–492]	443 [149–1317]	641 [360–1124]	0.12	0.19	0.88
	Non-O	A	275 [152–495]	172 [62–476]	477 [96.1–2375]	284 [144–558]	0.33	0.20	0.42
		B	533 [267–1064]	337 [112–1014]	185 [2.20–15,628]	979 [348–2752]	—	—	—
		AB	936 [447–1962]	876 [–]	1215 [0.0–1.7e09]	855 [302–2419]	0.075	0.46	0.052
Prevalence of clinical Pf cases % (number positive / total)	O		1 (0.5)	0 (0)	0 (0)	1 (1.6)	—	—	—
	Non-O		3 (0.9)	0 (0.0)	1 (0.0)	1 (1.8)	—	—	—
	Non-O	A	1 (0.7)	0 (0.0)	0 (0.0)	1 (2.6)	—	—	—
		B	1 (0.0)	0 (0.0)	0 (0.0)	1 (0.0)	—	—	—
		AB	2 (5.0)	0 (0)	1 (9.1)	1 (4.8)	—	—	—

F= Fulani; M= Mossi; R= Rimaibe; No-O = A + B + AB.

Table 4.
Malariometric indices according to the ethnicity and blood group/season.

4.4 Malaria infection and ABO blood groups

The subjects with Non-O blood (i.e. A, B, or AB) were less susceptible to malaria infection (p = 0.011). We found also a significant difference when Non-O groups were considered separately (O versus A, p = 0.028; O versus B, p = 0.04; O versus AB, p = 0.0067). **Table 5** shows malaria infection according to the blood group. Parasitological data did not differ when comparing subjects with and without blood group O, the most prevalent blood group in the population sample.

4.5 Malaria infection, ABO blood groups, and ethnicity

There was an association between the Non-O blood group of all ethnicities and malaria infections during high transmission. However, this association disappeared when the ethnic groups were considered separately (all ethnicities p = 0.011; Mossi: p = 0.56; Fulani: p = 0.59; Rimaibe: p = 0.49) (**Table 5**). Likewise, in the low transmission season, the difference in malaria infection between subjects with and without blood group O was not statistically significant (all ethnicities p = 0.86 Mossi: p = 0.72, Fulani: p = 0.94, Rimaibe: p = 0.91 On the other hand, a weak association between the Fulani and the Mossi with the

Blood group compared	All	Mossi	Fulani	Rimaibe
High transmission season				
Prevalence of Pf malaria Infection % (number positive/ total)				
O vs. A	0.028	0.17	0.12	0.49
O vs. B	0.04	0.80	0.04	0.06
O vs. AB	0.0067	0.90	0.11	0.13
O vs. (A + B + AB)	0.011	0.56	0.59	0.49
Geometric mean Parasite density (p/µl)				
O vs. A	0.74	0.98	0.47	0.30
O vs. B	0.91	0.34	0.18	0.57
O vs. AB	0.53	0.71	0.52	0.33
O vs. (A + B + AB)	0.000	0.58	0.52	0.29
Low transmission season				
Prevalence of Pf malaria Infection % (number positive/ total)				
O vs. A	0.79	0.90	—	0.80
O vs. B	0.84	0.65	—	0.87
O vs. AB	0.85	—	—	0.73
O vs. (A + B + AB)	0.86	0.72	0.94	0.91
Geometric mean (Parasite density (p/µl))				
O vs. A	0.55	0.88	0.59	0.41
O vs. B	0.28	0.34	0.53	0.27
O vs. AB	0.08	—	0.27	0.47
O vs. (A + B + AB)	0.55	0.58	0.69	0.09

Table 5.
The p values for the frequency of O and non-O blood group types between the three ethnic groups according to malaria infection and parasite density.

antigen of blood group O and the prevalence of asymptomatic malaria was found (61.7% with the Mossi against 45.5% with the Fulani; $p = 0.05$). Considering the subjects of blood group B, the study showed an association between Fulani versus Mossi/Rimaibe, $p = 0.001$; Fulani versus Mossi, P $p = 0.02$ and Fulani versus Rimaibe, $p = 0.003$.

5. Discussion

This study revealed that blood group O represented the most observed phenotype among participants (40.5%), followed by blood group B (26.6%), then blood group A (25.5%), and finally the AB blood group, which is rarer (7.3%). These results are consistent with those of previous studies that reported a high frequency of group O phenotypes in areas endemic to malaria [5, 9–13].

Other studies have corroborated this evidence by the inverse relationship: they have reported a high prevalence of blood group A and a low prevalence of phenotypes of blood group O in colder regions where malaria is an exotic disease [5, 10, 14].

These results would confirm the hypothesis of a selective evolutionary advantage (survival) of *Pf* infection on blood group O cells compared to other blood groups (A, B, or AB) in malaria-endemic areas [4].

The results of this study according to ethnicity also showed that Fulani were less infected with malaria despite their way of life and living in the same conditions of hyperendemic transmission. These were comparable to observations in previous studies in Burkina Faso and West Africa [15, 16].

Like the study carried out in Gabon by Monbo et al. [17], this study showed a high prevalence of malaria infection in participants with blood group O compared to participants with blood groups A, B, or AB; and is in contrast to previous studies that suggested individuals of blood groups A, B, and AB is more susceptible to *Pf* infection than group O [6, 18].

Lower parasite densities in subjects of blood group O compared to non-O subjects were observed, but the differences were not statistically significant. These observations are in line with the conclusions of previous studies which have shown that patients with blood group O were associated with increased protection against parasitemia [18–23].

These results would suggest a protective effect of the O antigen against clinical malaria. However, other explanations, such as the anti-rosette formation effect associated with blood group antigens, should also be considered [24].

In any case, this study made it possible to establish the involvement of Fulani groups against malaria infection and parasitemia. It also revealed that for all ethnicities combined, there would be a statistically significant difference in susceptibility to malaria between participants of blood groups O and participants of other blood groups (A, B, and AB).

The absence of any statistically significant difference in susceptibility to malaria in an intra-ethnic blood group analysis could be due to a sample size effect. Hence, the need for in-depth and broader epigenetic studies to accurately capture the effects of ABO blood groups in susceptibility to Pf malaria is highlighted.

6. Conclusion

The study confirmed that the Fulani group is less susceptible to *Pf* clinical malaria and when infected had generally lower parasite densities. The study also found that despite infectivity being more frequent in blood group O, individuals of this blood group are at less risk of clinical malaria and have low parasitaemias compared to individuals of other blood groups (A, B, or AB).

Evidence of correlations between ethnicity and blood group at risk of malaria infection would support the idea that the presumed association between blood group and malaria infection depends on the demographic distribution and characteristics of the population studied. As a result, each region of the world has a characteristic ABO phenotypic distribution, making it urgent to fully understand the biology of malaria infection through detailed studies of the interactions between the ABO blood grouping system: A critical condition for saving lives in malaria-endemic regions.

Authorship contributions

BEC wrote the manuscript and agree to be accountable for all aspects of the work.

AO and SBS read and approved the final manuscript and agree to be accountable for all aspects of the work.

Conflicts of interest

The authors have declared no conflict of interest.

Ethics approval

The study received approval from the ethical committees of the Ministry of Health of Burkina Faso.

Acronyms and abbreviations

GRAS	Groupe de Recherche Action en Santé
ABO	Group Sanguin (O, A, B, AB)
RBC	Red Blood Cells
Plasmodium falciparum	*P. falciparum*
WHO	World Health Organization

Author details

Edith Christiane Bougouma*, Alphonse Ouedraogo and Sodiomon Bienvenu Sirima
Groupe de Recherche Action en Santé (GRAS), Ouagadougou, Burkina Faso

*Address all correspondence to: bougouma.cedith@gmail.bf

References

[1] Calafell F, Roubinet F, Ramirez-Soriano A, Saitou N, Bertranpetit J, Blancher A. Evolutionary dynamics of the human ABO gene. Human Genetics. 2008;**124**:123-135

[2] Yamamoto F, Clausen H, White T, Marken J, Hakomori S. Molecular genetic basis of the histo-blood group ABO system. Nature. 1990;**345**:229-233

[3] Yamamoto F, Hakomori S. Sugar-nucleotide donor specificity of histo-blood group A and B transferases is based on amino acid substitutions. The Journal of Biological Chemistry. 1990;**265**:19257-19262

[4] Cserti CM, Dzik WH. The ABO blood group system and Plasmodium falciparum malaria. Blood. 2007;**110**: 2250-2258

[5] Uneke CJ. Plasmodium falciparum malaria and ABO blood group: Is there any relationship? Parasitology Research. 2007;**100**:759-765

[6] Fry AE, Griffiths MJ, Auburn S, Diakite M, Forton JT, Green A, et al. Common variation in the ABO glycosyltransferase is associated with susceptibility to severe Plasmodium falciparum malaria. Human Molecular Genetics. 2008;**17**:567-576

[7] Rowe JA, Handel IG, Thera MA, Deans AM, Lyke KE, Kone A, et al. Blood group O protects against severe Plasmodium falciparum malaria through the mechanism of reduced rosetting. Proceedings of the National Academy of Sciences of the United States of America. 2007;**104**: 17471-17476

[8] Kermarrec N, Roubinet F, Apoil PA, Blancher A. Comparison of allele O sequences of the human and non-human primate ABO system. Immunogenetics. 1999;**49**:517-526

[9] Dolo A, Modiano D, Maiga B, Daou M, Dolo G, Guindo H, et al. Difference in susceptibility to malaria between two sympatric ethnic groups in Mali. The American Journal of Tropical Medicine and Hygiene. 2005;**72**:243-248

[10] Zerihun T, Degarege A, Erko B. Association of ABO blood group and Plasmodium falciparum malaria in Dore Bafeno Area, Southern Ethiopia. Asian Pacific Journal of Tropical Biomedicine. 2011;**1**:289-294

[11] Seyoum S, Dagne K. ABO and rhesus blood-type frequencies in data from hospitals and the Red Cross in Ethiopia. Ethiopian Medical Journal. 1985;**23**:1-6

[12] Moges F, Kassu A, Mengistu G, Adugna S, Andualem B, Nishikawa T, et al. Seroprevalence of Helicobacter pylori in dyspeptic patients and its relationship with HIV infection, ABO blood groups and life style in a university hospital, Northwest Ethiopia. World Journal of Gastroenterology. 2006;**12**:1957-1961

[13] Hroob AMA, Saghir SAM, Almaiman AA, Alsalahi OSA, Al-Wajeeh AS, Al-Shargi OYA, et al. Prevalence and association of transfusion transmitted infections with ABO and Rh blood groups among blood donors at the National Blood Bank, Amman, Jordan. Medicina (Kaunas, Lithuania). 2020;**56**(12):701

[14] Godber M, Kopec AC, Mourant AE, Teesdale P, Tills D, Weiner JS, et al. The blood groups, serum groups, red-cell isoenzymes and haemoglobins of the Sandawe and Nyaturu of Tanzania. Annals of Human Biology. 1976;**3**:463-473

[15] Modiano D, Petrarca V, Sirima BS, Nebie I, Diallo D, Esposito F, et al. Different response to Plasmodium falciparum malaria in west African

sympatric ethnic groups. Proceedings of the National Academy of Sciences of the United States of America. 1996;**93**: 13206-13211

[16] Greenwood BM, Bradley AK, Greenwood AM, Byass P, Jammeh K, Marsh K, et al. Mortality and morbidity from malaria among children in a rural area of The Gambia, West Africa. Transactions of the Royal Society of Tropical Medicine and Hygiene. 1987;**81**:478-486

[17] Mombo LE, Ntoumi F, Bisseye C, Ossari S, Lu CY, Nagel RL, et al. Human genetic polymorphisms and asymptomatic Plasmodium falciparum malaria in Gabonese schoolchildren. The American Journal of Tropical Medicine and Hygiene. 2003;**68**:186-190

[18] Singh N, Shukla MM, Uniyal VP, Sharma VP. ABO blood groups among malaria cases from district Mandla, Madhya Pradesh. Indian Journal of Malariology. 1995;**32**:59-63

[19] Fischer PR, Boone P. Short report: Severe malaria associated with blood group. The American Journal of Tropical Medicine and Hygiene. 1998;**58**:122-123

[20] Lell B, May J, Schmidt-Ott RJ, Lehman LG, Luckner D, Greve B, et al. The role of red blood cell polymorphisms in resistance and susceptibility to malaria. Clinical Infectious Diseases. 1999;**28**:794-799

[21] Pathirana SL, Alles HK, Bandara S, Phone-Kyaw M, Perera MK, Wickremasinghe AR, et al. ABO-blood-group types and protection against severe, Plasmodium falciparum malaria. Annals of Tropical Medicine and Parasitology. 2005;**99**:119-124

[22] Montoya F, Restrepo M, Montoya AE, Rojas W. Blood groups and malaria. Revista do Instituto de Medicina Tropical de São Paulo. 1994;**36**:33-38

[23] Migot-Nabias F, Mombo LE, Luty AJ, Dubois B, Nabias R, Bisseye C, et al. Human genetic factors related to susceptibility to mild malaria in Gabon. Genes and Immunity. 2000;**1**:435-441

[24] Barragan A, Kremsner PG, Wahlgren M, Carlson J. Blood group A antigen is a coreceptor in Plasmodium falciparum rosetting. Infection and Immunity. 2000;**68**:2971-2975

Chapter 9

ABO Blood Groups and Risk of Glioma

Ana Azanjac Arsic

Abstract

Gliomas are one of the most common primary brain tumors and the etiology of gliomas remains unknown in most cases. The aim of this case–control study was to investigate possible association between incidence in relation to glioma and certain blood groups. This study included 100 histopathologically verified cases of glioma and 200 age and sex-matched controls without malignant diseases that were admitted to the same hospital. The results revealed that the patients with group AB were at 3.5-fold increased risk of developing glioma compared to the patients with other ABO blood groups. In this particular study, there was more male patients with glioma with the blood group AB. However, mechanisms that explain the relationship between the blood groups ABO and a cancer risk are unclear. Several hypotheses have been proposed, including the one with a modulatory role of blood group ABO antigens. In addition, the blood group ABO system regulates the level of circulating proinflammatory and adhesion molecules which play a significant role in the tumorigenesis process. Additionally, the recent discovery that includes the von Willebrand factor (vWF) as an important modulator of angiogenesis and apoptosis provides one plausible explanation as regards the role of the blood group ABO in the tumorigenesis process. To our knowledge, this is the first study that examined the relationship of blood group in patients diagnosed with glioma among the Serbian population. Moreover, for the first time our study results suggested that blood group AB increased the risk of glioma. The results of this study suggested that the blood group AB could be one of hereditary factors which had an influence on the occurrence of glioma. The further research is needed on a larger sample, to confirm these findings and the possible mechanisms by which the ABO system contributes to the pathology of glioma.

Keywords: ABO blood groups, risk factors, glioma

1. Introduction

The three most prevalent primary tumors of the brain, which represent the most common neoplasm of all brain tumors, are glioma, meningioma and pituitary adenoma. The most common primary brain tumors are gliomas [1]. Gliomas account for 27% of all central nervous system tumors and about 80% of malignant brain tumors. Based on the histological criteria, gliomas are classified into the following subtypes: astrocytoma, glioblastoma, oligodendroglioma, ependymoma, mixed glioma, malignant glioma, not otherwise specified (NOS) and a few rare histologies [2]. Each year, approximately 100,000 people are diagnosed with glioma worldwide. In addition, gliomas comprise less than 2% of all newly diagnosed cancers and are associated with substantial mortality and morbidity [3]. Glioblastoma multiforme (GBM) is the highest grade glioma (grade IV) tumor according to the grading system of the World Health Organization (WHO classification), and accounts for more than 50% of all gliomas [4]. The median overall survival of patients with glioma is approximately 14-17 months, whereas the incidence of gliomas is 6.8/100,000 and it is increasing worldwide. The countries with Northern European populations had higher incidence rates (ranging from 7.8 in the USA to 9.6 in Australia and New Zealand). However, a decline in the incidence rates was noticed in countries with predominantly Asian and Africans populations (ranging from 1.9 in Southeast Asia to 3.3 in India) [5]. Considering the fact that the etiology of glioma is largely unknown, numerous risk factors have been examined as potential contributors to glioma risk. As a result, several potential factors are found to be increasing the risk of developing glioma and they included environmental factors such as smoking, alcohol consumption, diet, obesity, infections, environmental pollution, and ionizing radiation [6, 7]. In this regard, some epidemiological studies suggested that allergic conditions (such as asthma, hay fever, eczema, food allergies, etc.) reduced the risk of glioma [8]. Well-established risk factors for glioma development include older age, male gender, Caucasian race/ethnicity, complete with rare genetic syndromes. Li-Fraumeni syndrome, neurofibromatosis (types 1 and 2), tuberous sclerosis, nevoid basal cell carcinoma syndrome, familial adenomatous polyposis (FAP), and von Hippel–Lindau (VHL) syndrome cause a small percentage of gliomas in adult populations [9–11]. In addition, ABO blood groups, expressed by different cells in human tissues, including epithelial cells, vascular endothelial cells, and neurons [12], have been associated with an increased cancer risk.

Many recent studies have focused on the role of ABO blood group antigens in the pathogenesis of various systemic diseases, including cancer. Recently, more evidence suggested that there was a significant association between the distribution of ABO blood group antigens and a risk of tumors, including the development of pancreatic cancer, hepatocellular carcinoma, nasopharyngeal carcinoma, and gastric cancer [13–15].

Furthermore, several studies have shown conflicting results concerning the relationship between ABO blood groups and glioma risk [16–29]. Finally, differences are observed not only in the distribution of ABO blood group antigens among patients with primary brain tumors in various countries worldwide, but also among different ethnic groups within the country. The influence of blood group types on the pathogenesis of brain tumors is still unclear, considering the fact that there are conflicting reports obtained from the studies related to the distribution of ABO blood groups and a risk of developing glioma.

2. The relationship between ABO blood group and cancer risk

The ABO blood group system was discovered at the University of Vienna in 1901 by Austrian scientist, Karl Landsteiner. The locus for the ABO blood group is on chromosome 9, whereas ABO gene is inherited in an autosomal dominant [30]. The A and B antigens are displayed on the surface membrane of red blood cells and attached to a common precursor of the side chain, the H determinant, which is then converted into the A or B antigen. As regards the expression of antigens, individuals with blood type AB have both antigens (A, B). On the other hand, individuals with blood type A have antigen A, individuals with blood type B have antigen B, whereas individuals with blood type O have neither antigen A, nor antigen B. Moreover, the individuals of group O lack such functional enzymes and express the unaltered H determinants. In addition to their expression on the erythrocytes, A and B antigens are highly expressed on the surface of epithelial cells of gastrointestinal, broncho-pulmonary, and urogenital tracts. Ultimately, they are expressed on the surface of the neurons, platelets, and vascular endothelium [31, 32].

The possibility of association between ABO blood type and malignancy was first explored by Anderson and Haas. ABO blood group antigens are widely expressed in a variety of human cells and tissues. As previously mentioned, the underlying mechanisms that could explain the direct relationship between the ABO blood groups and cancer risk still remains a challenge. One of the probable explanations refers to the modulatory role of ABO blood group antigens. In addition, a variety of tumors show modified expression of ABO antigens on the surface of cancer cells in comparison with normal epithelial cells [32–34]. Moreover, what may prevent the immune system from recognizing and destroying tumor cells is the structural similarity observed between ABO blood groups antigens and tumor antigens. That results in a greater risk of cancer development and progression [35]. In particular, glycoconjugates are considered to be the key mediators of membrane **signaling** and intracellular adhesion. Hence, they are necessary for malignant progression and metastasis [36]. In addition, a link was detected between ABO gene polymorphism and proinflammatory and adhesion molecules (such as E-selectin, P-selectin, and intracellular adhesion molecule-1) which play an important role in tumorigenesis. It is well known that low levels of intercellular adhesion molecule –1 (ICAM-1) facilitate the adhesion of some cancer cells to endothelial cells in patients with a non-O blood group. Furthermore, low-level ICAM-1 may accelerate cancer progression [37].

Taking into account that the Von Willebrand factor (vWF) plays an indirect role in the tumorigenesis process, it has to be emphasized that its main function is to initiate platelet adhesion to the endothelial cells upon vascular injury. Being an important modulator of angiogenesis and apoptosis, vWF is secreted by vascular endothelial cells. The lowest serum levels of these factors have been associated with the O blood group, the intermediate levels – with blood groups A and B, whereas the highest levels have been associated with blood group AB. In this manner, we did observe a positive correlation between levels of vWF and disease severity. The level of vWF is an indicator of the extent of endothelial dysfunction caused by tumor growth. Thus, elevations in the concentrations of plasma vWF could be related to accelerated tumor growth in individuals with non-O blood groups [38, 39]. As regards the Forssman antigen (FORS1 Ag), which is structurally similar to the structure of A antigen determinant of the ABO blood group system, it is synthesized predominantly in stomach and colon cancer. However, one study confirmed the presence of Forssman antigen in hepatocellular carcinoma tissues [40]. Alterations in surface antigens, particularly glycoconjugates, may not only lead to modifications in intercellular adhesion, but could have an important role in the development of cancer as well. There is still no formal hypothesis which would provide a plausible explanation about the association between malignancy and ABO blood groups [35]. Consequently, the ABO gene may be in linkage disequilibrium with other genes influencing cancer risk [41].

2.1 The relationship between ABO blood groups and risk of pancreatic cancer

The results obtained from three previously conducted studies indicated a lower incidence of pancreatic cancer among people with blood group O. Additionally, the Nurses' Health Study and the Health Professionals Follow-Up Study demonstrated an increased incidence of pancreatic cancer among subjects with blood type antigen A or B compared with those who were lack of these antigens [42–44].

Dandona et al., also observed an increased risk of developing pancreatic cancer in patients with a non-O blood group [45].

2.2 The relationship between ABO blood group and risk of gastric cancer

Based on the results of previously conducted study, which included 3,623 patients from England and Scotland, a significantly higher frequency of the A blood group and a lower frequency of the O blood group was observed among stomach cancer patients. In addition to the above mentioned findings, it was also revealed that the people with blood group B showed a significantly reduced risk of stomach cancer [46]. On the other hand, several studies, mostly conducted in Western populations, consistently showed an approximately 20% excess risk of gastric cancer in individuals with blood type A [47]. The results obtained from this particular study confirmed in a recent meta-analysis [48]. Furthermore, there was a prospective cohort study in a Taiwanese population that showed the fact that blood type A was associated with a 38% increased risk of stomach cancer [49]. In addition, another prospective cohort study that included middle-aged or older Chinese men demonstrated a lower risk of all cancer for blood type B, as well as a lower risk for gastrointestinal cancers including stomach and colorectal cancer for blood types B and AB rather than blood type A [50]. In a study by Aird et al., no association was found between the ABO blood group and risk of stomach cancer [51].

2.3 The relationship between ABO blood groups and lung cancer

The increased risk of developing lung cancer was observed in patients with non-O blood type in Turkish populations [52].

2.4 The relationship between ABO blood groups and breast cancer

Having studied the association between ABO blood groups and breast cancer, Stamatokos et al., suggested that an A antigen was associated with an increased risk of developing breast cancer. In contrast, Tryggvadottir and colleagues observed an increased risk among women with B antigen [53, 54]. On the basis of the results of a meta-analysis, it was concluded that blood group A was positively linked with an increased risk of breast cancer [55].

2.5 The relationship between ABO blood groups and colon cancer

According to the findings of a meta-analysis, which evaluated the association between the ABO blood group and colon cancer, the protective effects of O blood type were confirmed. However, blood types A and B were not linked to a higher risk of colon cancer [56].

2.6 The influence of ABO blood group on the development of glioma

The Central Nervous System (CNS) lesions are known to mankind since 1774. Systemic study of the CNS commenced when Baily and Cushing started their studies in the early 1920s [57]. The association between brain tumors and blood groups antigens is variable. No specific hypothesis has been proposed for the association between CNS neoplasms and ABO blood groups. The abovementioned mechanisms may operate even in the CNS neoplasms. Kumarguru et al., suggested a probable hypothesis which was related to the fact that an alteration in the characteristics of ABO blood group antigens on the surface of cell origin, under the influence of either environmental factors or genetic factors, may govern the process of development of tumors. In this regard, the mechanism may operate not only in the primary neoplastic lesions but in the metastatic lesions of the CNS in the genetically susceptible individuals as well [58]. Taking into account that there are conflicting reports obtained from studies done on the distribution of ABO blood groups in primary intracranial neoplasms, the influence of blood group types on the pathogenesis of brain tumors still remains unclear. Periayavan S et al., observed that blood group O was common in most of the categories of CNS lesions: neuroepithelial tumors (38.45%), meningeal (37.57%), cranial and paraspinal nerve tumors (39.67%), pituitary neoplasms (43.62%), and metastatic tumors (43.18%) [59]. As regards the study undertaken by Mehrazin M et al., neuroepithelial tumors (38.4%), cranial and paraspinal nerve tumors (35.9%), and pituitary neoplasms (4.40%) were observed more frequently in O blood group patients in their study [59]. In contrast, previous studies showed different results concerning the ratios of blood groups among the patients with glioma. Yates and Pearce conducted the first study which examined the relationship between the ABO blood group and the risk of glioma [16]. In this respect, there was no significant relationship found between ABO blood groups and risk of glioma among patients diagnosed before 1945. After that year, a highly statistically significant decrease in number of patients with blood group O was reported. The relationship between the blood groups and astrocytic brain tumors was examined in the study conducted by Selvestrone and Cooper [17], the results of which showed a statistically significant decrease in the number of patients with blood group B and O. In addition to the abovementioned studies, in an Italian case–control study that recruited 195 cases of histologically-confirmed glioma, a positive association was found with blood group A when low-grade astrocytomas were considered separately. Additionally, the present study identified suggestive, but non-statistically significant association with the presence of CNS tumors among the first and second-degree relatives [18]. Compared to the general population, Campbell et al., reported a substantially higher incidence of glioma in blood group O individuals [19]. Yates et al., did not detect any significant differences in the distribution of blood group antigens between glioma patients and controls from the Oxford region in the United Kingdom [20]. These results were confirmed by Strang et al., in a study that included 900 astrocytoma patients and the control population [21]. Furthermore, there was a significantly higher number of male cerebral astrocytoma patients with group A confirmed in this particular study.

The retrospective study conducted by Mehrazin et al., was made on 907 histologically confirmed cases of glioma and the same number of age and sex-matched controls. The distribution of ABO blood groups in this study population was compared with that of the general population. Finally, the results showed no significant differences between types of intracranial tumors and frequencies of four blood groups [22]. The case–control study by Akhtar et al., on 112 central nervous system tumors, found a significantly higher association of these tumors with blood group B patients [23]. On the contrary, in a case–control study, Akca et al., aimed to compare the patients with glioblastoma multiforme with control groups regarding ABO blood groups. On the basis of the study results, no significant differences among O, A, B and AB blood groups were shown [24]. In a study conducted by Turowski and Czochra, the analysis was based on the distribution of ABO blood groups in 271 patients treated for glioblastoma multiforme (GBM), whereas the control group included 500 patents with craniocerebral trauma. Hence, a statistically significant difference was found in the distribution of ABO blood groups between these patient groups. Moreover, higher frequency of group A and lower frequency of group O was detected in GMB patients [25]. In the cohort study, Allouh et al., included

115 patients who were diagnosed with glioblastoma multiforme (GB) in Jordan, between 2004 and 2015. Inclusion criteria were histologically confirmed glioma. Due to hospital records, data related to patients' characteristics (such as the following: age, sex, ABO blood groups, and Rh factor) were collected. Consequently, the study results suggested that individuals with group A had a higher than expected chance of developing glioblastoma multiforme, while individuals with group O had a lower chance. In addition, a lower incidence of glioblastoma multiforme was reported in individuals with group O compared to healthy blood donors and age- and sex-matched control subjects [26]. The case–control study in the Serbian population, which consisted of 100 pathohistologically confirmed glioma and 200 age and sex-matched control individuals, observed a higher incidence of glioma in patients with the blood group AB [27].

The aim of the retrospective study done by Chang et al., was to examine the relationship between ABO blood groups and brain tumors in the Chinese population. This study included 2077 cases with histologically confirmed glioma admitted at the hospital in the period between 2001 and 2016, whereas the control group included 2716 noncancer patients admitted to the same hospital. Consequently, the results showed that blood types B and AB were significantly linked to the risk of glioma [28]. In a prospective study, involving more than 100,000 adults in the United States and nearly 20 years of follow-up, no statistically significant differences in risk of glioma were identified by ABO blood type. Hence, the abovementioned study suggested that ABO blood group may not play a role in the development of glioma [29].

3. Prognostic value of ABO blood groups in patients with glioma

Some studies investigated the prognostic value of ABO blood groups in patients with glioma. Alkan et al., performed a retrospective cross-sectional study which included 759 patients with glioma. In this study, there was no statistically significant difference observed between glioma patients and healthy control patients. Median overall survival (mOS) of GBM patients were 12.9 months in A, 13.4 months in B, 5.7 months in AB, 12.8 months in 0 blood groups. The median overall survival of anaplastic astrocytoma patients were 24.4 months in A, 47.2 months in B, 37.8 months in AB, 29.2 months in 0 blood groups respectively. Therefore, these results showed that ABO blood groups had no prognostic value [60]. In a retrospective study on 72 patients with GMB, Akca et al., found no correlation among ABO blood groups and prognosis. In that study, the number of patients with blood groups O, A, B, and AB were 23, 33, 9 and 7, respectively [24]. In a retrospective observational study, Sokmen et Karcin showed that overall survival was shorter in GMB patients with non-O ABO blood groups than those with blood group O. This study consisted of 238 patients with GMB and it was the first study showing the prognostic value of ABO blood groups in patients with GMB [61]. The reason may be that the number of patients was insufficient for prognostic analyses in the study conducted by Akca et al.

The previous study, which aimed to evaluate the prognostic value in overall survival, had some limitations. With this respect, some of the studies failed to obtain data regarding O^6-methylguanine-DNA methyltransferase (MGMT) promoters, which are shown to have prognostic value in GMB. Additionally, some studies suggested that overall survival be shorter in GMB patients with non-O ABO blood groups than those with blood group O.

It is confirmed that differences in the distribution of ABO among patients with primary brain tumors have been reported between countries worldwide, complete with differences between ethnic groups within the country which are observed as well.

As regards the effects of ABO blood groups on the incidence and survival of patients with glioma, they are still unknown. What can be confirmed is that ABO blood status has been shown to be associated with many cancers. However, the influence of blood group types on the pathogenesis of glioma is still unclear. Due to the difference of ABO genes among socioeconomic groups and geographic areas, the blood groups may have a role in the incidence and prognosis of glioma just like the environmental factors as well.

4. Conclusion

In conclusion, it should be stated that gliomas are considered to be the most common primary brain tumors, accounting for almost 80% of all malignant brain tumors. The etiology of glioma is still largely unknown. However, in the past decades, a lot of research has been focused on determining risk factors that may contribute to glioma aetiopathogenesis, including genetic and environmental factors. Previous reports demonstrated that gene mutations and ABO blood groups may be associated with elevated glioma risk. Therefore, further research of such kind is needed to be performed on a larger sample, for the purpose of confirming the abovementioned findings complete with potential mechanisms by which the ABO system contributes to the pathology of glioma.

Author details

Ana Azanjac Arsic
University of Kragujevac, Serbia, Faculty of Medical Sciences, Department of Neurology, Kragujevac, Serbia

*Address all correspondence to: ana.azanjac@yahoo.com

References

[1] Ostrom QT, Bauchet L, Davis FG, Deltour I, Fisher JL, Langer CE, et al. The epidemiology of glioma in adults: A "state of the science"review. Neuro Oncol. 2014;16(7):896-913.

[2] Ostrom QT, Gittleman H, Fulop J, Liu M, Blanda R, Kromer C, et al. CBTRUS Statistical Report: Primary Brain and Central Nervous System Tumors Diagnosed in the United States in 2008-2012. Neuro Oncol. 2015;17(4):iv1–iv62.

[3] Ferlay J, Colombet M, Soerjomataram I, Mathers C, Parkin DM, Piñeroset M, et al. Estimating the global cancer incidence and mortality in 2018: GLOBOCAN sources and methods. Int J Cancer. 2019;144:1941-1953.

[4] Jhanwar-Uiyal M, Labagnara M, Friedman M, Kwasnicki A, Murali R. Glioblastoma molecular pathways, stem sells and therapeutic targets. Cancer (Basel). 2015;7:538-555.

[5] Leece R Xu J, Ostrom T, Chen Y, Kruchko C, Barnholtz-Sloan JS. Global incidence of malignant brain and other central nervous system tumors by histology, 2003-2007. Neuro Oncol. 2017; 19:1553-1564.

[6] Malmer B, Adatto P, Armstrong G, Barnholtz-Sloan J, Bernstein JL, Claus E, et al. GLIOGENE an International Consortium to Understand Familial Glioma. Cancer Epidemiol Biomarkers Prev. 2007;16(9):1730-1734.

[7] Bondy ML, Scheurer ME, Malmer B, Barnholtz-Sloan JS, Davis FG, Il'yasova D, et al. Brain tumor epidemiology: consensus from the Brain Tumor Epidemiology Consortium. Cancer. 2008;113(7):1953-1968.

[8] Berg-Beckhoff G, Schüz J, Blettner M, Münster E, Schlaefer K, Wahrendorf J, Schlehoferet B. History of allergic disease and epilepsy and risk of glioma and meningioma (INTERPHONE study group, Germany). Eur J Epidemiol. 2009;24(8):433-440.

[9] Inskip PD, Linet MS, Heineman EF. Etiology of brain tumors in adults. Epidemiol Rev. 1995;17(2):382-414.

[10] Bondy M, Wiencke J, Wrensch M, Kyritsis AP. Genetics of primary brain tumors: a review. J Neurooncol. 1994;18(1):69-81.

[11] Preston-Martin S and Mack WJ. Neoplasms of the nervous system. InD. Schottenfeld and J.F. Fraumeni, Jr. (eds.), Cancer epidemiology and prevention (2nd ed.), p. 1231-1291, Oxford University Press, New York (1996).

[12] Franchini M, Liumbruno GM. ABO blood group and neurodegenerative disorders: more than a casual association. Blood Transfus. 2016;14:158-159.

[13] Li Q, Yu CH, Yu JH, Liu L, Xie SS, Li WW, et al. ABO blood group and the risk of hepatocellular carcinoma: A casecontrol study in patients with chronic hepatitis B. PLoS One.2012; 7(1):e29928.

[14] Song HR, Shin MH, Kim HN, Piao JM, Choi JS, Hwang JE, et al. Sex specific differences in the association between ABO genotype and gastric cancer risk in a Korean population. Gastric Cancer.2013; 16(2):254-260.

[15] Rizzato C, Campa D, Pezzilli R, Soucek P, Greenhalf W, Capurso G, et al. ABO blood groups and pancreatic cancer risk and survival: Results from the PANcreatic Disease ReseArch (PANDoRA) consortium. Oncol Rep. 2013;29(4):1637-1644.

[16] Yates PO, Pearce KM. Recent change in blood-group distribution of astrocytomas. Lancet. 1960;1(7117): 194-195.

[17] Selvestrone B, Cooper DR. ABO blood group and astrocytomas. JNeurosurg. 1961;18:602-604.

[18] Zampieri P, Meneghini F, Grigoletto F, Gerosa M, Licata C, Casentini L, et al. Risk factors for cerebral glioma in adults: A casecontrol study in an Italian population. J Neurooncol. 1994;19(1):61-67.

[19] Campbell ACP, Gaisford W, Patterson E and Steword JK. Tumors in children: A survey carried out in the Manchester region. Br Med J. 1961;1(5224):448-452.

[20] Yates PO. Malignant glioma and ABO blood group. Br Med J. 1964;1(5378):310.

[21] Strang RR, Tovi D, Lopez J. Astrocytomas and the ABO blood groups. J Med Genet. 1966;3:274-275.

[22] Mehrazin M. ABO blood group frequency and brain tumors. Asian Pac J Cancer Prev. 2006;7:582-584.

[23] Akhtar K, Mehdi G, Sherwani R, Sofi L. Relationship between various cancers and ABO blood groups: A Northern India experience. Int J Pathol. 2010;13(1):1-4.

[24] Akca Z, Mutlu H, Erden A, Buyukcelik A, Sezer Y, Inal A. The relationship between ABO blood group and glioblastoma multiforme. Med Sci. 2014;3(4):1639-1647.

[25] Turowski K, Czochra M. ABO blood groups in glioblastoma multiforme. Neurol Neurochir Pol. 1979;13:173-176.

[26] Allouh MZ, Barbarawi AM, Hiasat MY, Al-Qaralleh MA, AbabnehEI. Glioblastoma and ABO blood groups: Further evidenceof an association between the distribution of blood group antigens and brain tumors. Blood Transfus. 2017;15(6): 543-547.

[27] Azanjac Arsić A, Miletić Drakulić S, Vesić K, Tončev blood group and risk of glioma: a case control study from Serbia. Vojnosanit Pregl. 2019;76(5):465-469.

[28] Chang L, Zhao S, Dan Z, Li G, Zhang D, Tan C, et al. ABO blood group system and risk of brain tumors: A retrospective case–control study in Northeast of China. Tranl Srg. 2018;3:49-53.

[29] ABO blood group and risk of glioma. Letters to the editor. Neurooncology. 2017;19(6):871-874.

[30] Landsteiner K. Über Agglutinationserscheinungen normalenmenschlichen Blutes. In: Harper PS, editor. Landmarks inMedical Genetics. Oxford: Oxford University Press; 1901. p.

[31] Liumbruno GM, Franchini M. Beyond immunohaematology: Therole of the ABO blood group in human diseases. Blood Transfus. 2013;11(4): 491-499.

[32] Hsiao L, Liu N, You S, Hwang L. ABO blood group and the risk of cancer among middle-aged people in Taiwan. Asia Pac JClin Oncol. 2015;11(4):e31–e36.

[33] Jovanović SS, Veljković DK. Immunobiological and clinical significance of blood groups. Beograd: Intra Net Communication;2009. (Serbian).

[34] Louis DN, Perry A, Reifenberger G, Deimling A, Figarella-BrangerD, Cavenee WK, et al. The 2016 World Health Organization Classification of Tumors of the Central Nervous System: Asummary. Acta Neuropathol. 2016;131(6):803-820.

[35] Franchini M, Lippi G. The intriguing relationship between the ABO blood group, cardiovascular disease, and cancer. BMC Med. 2015;13:7.

[36] Cao X, Wen ZS, SunYJ, Li Y, Zhang L, Han YJ. Prognostic value of ABO blood group in patients with surgically resected colon cancer. British Journal of Cancer. 2014;111:174-180.

[37] Witkowska AM, Borawska MH. Soluble intercellular adhesion molecule-1 (sICAM-1): An overview. Eur Cytokine Netw. 2004;15(2):91-98.

[38] Jenkins PV, O'Donnell JS. ABO blood group determines plasma von Willebrand factor levels: a biologic function after all?. Transfusion. 2006;46:1836-1844.

[39] O'Donnell J, Laffan MA. The relationship between ABO histoblood group, factor VIII and von Willebrand factor. Transfus Med. 2001;11:343-351.

[40] Prieto PA, Smith DF. A new ganglioside in human meconium defected by antiserum against the human milk sialylohigosaccharide 2S-tetrasaccharide b. Archives of Biochemistry and Biophysics. 1985;241(1):281-289.

[41] Vogel F. Controversy in human genetics. ABO blood groups and disease. Am J Hum Genet. 1970;22(4):464-475.

[42] Wolpin BM, Chan AT, Hartge P, Chanock SJ, Kraft P, Hunter DJ, et al. ABO blood group and the risk of pancreatic cancer. J Natl Cancer Inst. 2009;101:424-431.

[43] Wolpin BM, Kraft P, Gross M, Helzlsouer K, Bueno-de-Mesquita HB, Steplowski E, et al. Pancreatic cancer risk and ABO blood group alleles: results from the pancreatic cancer cohort consortium. Cancer Res. 2010;70: 1015-1023.

[44] Nakao M, Matsuo K, Hosono S, Ogata S, Ito H, Watanabe M, et al. ABO blood group alleles and the risk of pancreatic cancer in a Japanese population. Cancer Sci. 2011;102: 1076-1080.

[45] Dandona M, Gao F, Linehan DC, Wang-Gillam A. ABO blood group and the risk of pancreatic cancer. J Natl Cancer Inst. 2010;102:135-137.

[46] Aird I, Bentall HH, Mehigan JA, Roberts JA. The blood groups in relation to peptic ulceration and carcinoma of colon, rectum, breast, and bronchus; an association between the ABO groups and peptic ulceration. Br Med J. 1954; 2(4883):315-321.

[47] Newman E, Naifeh GS, Auer JE, Buckwalter JA. Secretion of ABH antigens in peptic ulceration and a gastric carcinoma. Br Med J. 1961;1(5219):92-94.

[48] Wang Z, Liu L, Ji J, Zhang J, Yan M, Zhang J, et al. ABO blood group system and gastric cancer: a case–control study and meta-analysis. Int J Mol Sci. 2012;13(10):13308-13321.

[49] Sun W, Wen CP, Lin J, Wen C, Pu X, Huang M, et al. ABO blood types and cancer risk. Đa cohort study a of 339,432 subjects in Taiwan. Cancer epidemiology. 2015;39(2):150-156.

[50] Huang JY, Wang R, Gao YT, Yuan YM. ABO blood type and the risk of cancer - Findings from the Shanghai Cohort Study. PLoS One. 2017;12(9):e0184295.

[51] Aird I, Bentall HH, Roberts JA. A relationship between cancer of stomach and the ABO blood groups. Br Med J. 1953;1(4814):799-801.

[52] Urun Y, Utkan G, Cangir AK, Oksuzoglu OB, Ozdemir N, Oztuna DG, et al. Association of ABO blood group and risk of lung cancer in a multicenter

study in Turkey. Asian Pac J Cancer Prev. 2013;14(5):2801-2803.

[53] Stamatakos M, Kontzoglou K, Safioleas P, Safioleas C, Manti C, Safioleas M. Breast cancer incidence in Greek women in relation to ABO blood groups and Rh factor. Int Semin Surg Oncol. 2009;6:14.

[54] Tryggvadottir L, Tulinius H, Robertson JM. Familial and sporadic breast cancer cases in Iceland: a comparison related to ABO blood groups and risk of bilateral breast cancer. Int J Cancer. 1988;42(4): 499-501.

[55] Miao SY, Zhou W, Chen L, Wang S, Liu XA. Influence of ABO blood group and Rhesus factor on breast cancer risk: a meta-analysis of 9665 breast cancer patients and 244,768 controls. Asia Pac J Clin Oncol. 2014;10(2):101-108.

[56] Cao X, Wen ZS, Sun YJ, Li Y, Zhang L, Han YJ. Prognostic value of ABO blood group in patients with surgically resected colon cancer. Br J Cancer. 2014;111(1):174-180.

[57] Bailey P, Cushing H. A classification of the tumors of the glioma group on a histogenetic basis with a correlated study of prognosis, JB Lippincott, Philadelphia 1926.

[58] Kumarguru BN, Pallavi P, Sunila, Manjunath GV, Vasan TS, Rajalakshmi BR. Histopathological study of central nervous system lesions: emphasizing association of neoplasms with ABO blood groups. Journal of clinical and diagnostic research. 2017;11(4):EC15-EC20.

[59] Periyavan S, Malini AV, Radha RK, Sangeetha SK, Subba Krishna DK, Shankar SK. Association of ABO blood group system to intracranial tumors-a study from South India. Indian J Pathol Microbiol. 2006;49:473-475.

[60] Alkan A, Yazıcı G, Cengiz M, Karakaş Y, Çelik İ, Kars A, et al. The role of ABO blood groups in glial neoplasms. Br JNeurosurg. 2019;33(1):43-46.

[61] Sokmen FC, Karacin C. Prognostic value of ABO blood groups in patients with glioblastoma multiforme. J Coll Physicians Surg Pak. 2020;30(7): 713-716.